The Controller as Lean Leader

A Novel on Changing Behavior with a Lean Cost Management System

Sue Elizabeth Sondergelt

CRC CRC Press
Taylor & Francis Group
Boca Raton London New York

CRC Press is an imprint of the
Taylor & Francis Group, an **informa** business
A PRODUCTIVITY PRESS BOOK

CRC Press
Taylor & Francis Group
6000 Broken Sound Parkway NW, Suite 300
Boca Raton, FL 33487-2742

© 2012 by Taylor & Francis Group, LLC
CRC Press is an imprint of Taylor & Francis Group, an Informa business

No claim to original U.S. Government works

Printed in the United States of America on acid-free paper
Version Date: 20120321

International Standard Book Number: 978-1-4398-8277-1 (Paperback)

Library of Congress Cataloging-in-Publication Data

Sondergelt, Sue Elizabeth.
    The controller as lean leader : a novel on changing behavior with a lean cost management system / Sue Elizabeth Sondergelt.
        p. cm.
    Includes bibliographical references and index.
    ISBN 978-1-4398-8277-1 (pbk. : alk. paper)
    1. Manufacturing industries--Accounting. 2. Cost accounting. 3. Managerial accounting. I. Title.

HF5686.M3S64 2012
657'.42--dc23   2012008927

Visit the Taylor & Francis Web site at
http://www.taylorandfrancis.com

and the CRC Press Web site at
http://www.crcpress.com

Contents

Preface

This book is a compilation of, and reflection upon, all of my Lean and Lean accounting experiences in businesses attempting Lean, with clients from my Lean Beans LLC consulting business, with universities where I have taught as an adjunct, as well as with friends whom I have helped and who have helped me. The story includes the good, the bad, the ugly … and the humorous! The story involves a fictional manufacturing entity that is embarking on a Lean change management journey—for the second time, having failed at its first attempt at Lean implementation just a few years earlier. What will the company do differently this time? And what will it do to ensure that it does not slip backward again as the transformation unfolds?

The characters, situations, locations, and organizations described in this novel are entirely fictional, as is any production process or HR model described herein. Any similarity between the characters in this book and actual people, situations, locations, and organizations is purely coincidental. The significance and value that one should take from the reading of this novel, however, are not fictional. Based on two decades of experiences in successful (as well as not so successful) attempts at Lean transformation and Lean accounting implementation, this book tells the story of what can be if you focus on the behavior, and changing the behavior, of people rather than on the numbers. The heart of this book is truly the people and a Lean Cost Management System to change behavior, which in turn will change the culture of any Lean enterprise.

The process of implementing Lean accounting, as described in this book, is not a recipe to be copied but, rather, the result of a series of experiments, trials, and research. It is people's initiative and problem solving that are to be admired. The end result is a learning organization with its people still engaged in learning to understand what works best for their business. The most important lessons to be learned here are threefold: have a vision and purpose with shared values, and live those values; take the initiative and do the research to find what works best for your organization; and, most of all, *be humble* and believe in yourself.

I wish to thank all of you, both management and my associates, in two businesses where I have worked, namely, The Wiremold Company and United Technologies–Pratt & Whitney, who contributed to my Lean learning over the years, as well as those of you who assisted me and supported me in the writing of this book. I especially want to thank a very dear friend outside of business who, together with her dogs Sarah and Charley, taught me so much more about behavior and motivation. I also wish to thank Michael Sinocchi, Senior Acquisitions Editor with Productivity Press, Taylor & Francis Group, for giving me the opportunity to tell my Lean story from the viewpoint of a financial person working in operations.

My purpose in writing this book is to bring to light some of the many misconceptions of Lean accounting as well as the various misconceptions about the implementation of the Lean cost management system; the importance of a cost management system as opposed to a cost accounting system, the controller's extended nonfinancial leadership role in this entire change management endeavor, and what should be the role of academia in business today. The main theme of the book is that culture, simply stated, is merely the sum of all behaviors in an entity, and thus to change the culture one must first change behavior, and the best way to change behavior is to get rid of the traditional cost accounting system that is creating all the wrong behavior and replace it with a Lean cost management system. A secondary theme of the book deals with the timing of the implementation of the various pieces of the Lean cost management system to ensure success in our Lean transformation. I do not profess to be an "expert" as I believe it was Henry Ford himself who told us that the minute a person becomes an "expert" is the time for that person to get out of business. I am simply telling my story, in an effort to help others. In writing this book, I was motivated and challenged to dig deeper to better understand the overall philosophy of business that Henry Ford probably called "common sense" and that Toyota today calls "TPS." The descriptions of activities and processes in this book are in no way meant to accurately portray either Ford's or Toyota's Lean models.

Author

Having begun her professional career as a high school mathematics teacher, **Sue Elizabeth Sondergelt** gained insight into something other than numbers, namely, soft skills and people skills—skills that she learned, upon entering business, many other financial professionals did not possess. After going back to school to earn an MS in accountancy and her CMA (Certified Management Accountant), she went to work for The Wiremold Company, a world-class organization and textbook case in Lean and Lean accounting. Then, after 15 years of service in operations and finance in two manufacturing conglomerates, Sue semi-retired, and returned to education as an adjunct at two universities, one online and one brick-and-mortar. It was here that she realized that the behavior we see in business today is "learned" behavior, and that it has been learned, both by engineers and by accountants, at our universities. Sue believes Lean accounting will never become a valid, well-utilized, and respected cost management system until we begin to teach it in our accounting texts and curricula, including our MBA programs, at all universities.

Since 2007, Sue has been the creator of the innovative new Lean Beans program on Lean accounting, and owner of Lean Beans LLC, an education and consulting group in Lean accounting. She has delivered her highly respected seminar, "Lean Thinking for Accountants: Sustaining Corporate Growth," for over three years and in 30 major U.S. cities, to assist CPAs in fulfilling their continuing education requirements in cost management. Today she spends most of her time assisting small manufacturers in their Lean cost management journey.

This new book, *The Controller as Lean Leader: A Novel on Changing Behavior with a Lean Cost Management System,* brings together the good and the bad from all of her experiences. She presents information in a way that will change the way you think. She creates surprising and compelling visuals to change the way you feel about finance in business today. She guides readers through the first three principles of Lean, which are so crucial to the successful implementation of the Lean cost management system. Then she leads the reader to understand how only a Lean direct cost management system will drive the correct "behavior" to sustain a Lean change management initiative and change the culture in any organization today.

1

Unbridled, Unproductive Chaos!

It is six in the morning, and I am rolling into the parking lot. I have always believed in getting an early start on the day, but perhaps not quite this early. Although I am new to my job at this plant (but not this corporation), I perceive that coming in at six in the morning will become a necessity. There seems to be so much to do here!

I was brought into this particular plant to help "raise the bar." By profession, I am a managerial accountant and controller. I like to qualify that and say that I am a Lean controller. I believe that the finance "function" plays an integral role in the Lean management revolution. I believe that effectiveness is more important than efficiency and that finance people can become more effective by getting to know the business "cold." And I believe that the most important role for the finance person is that of catalyst in a cultural change initiative. I also believe that a business should not focus so much on numbers but rather on behavior.

The chaos that is happening at this plant is not just the result of a lack of "financial systems" as the corporate internal audit group had told me, and it is not just about a "war" between operations (OPS) and accounting. Rather it is about the "learned behavior" of both operations and accounting associates. OPS people have been taught how to "play the game" in their MBA programs at the university. And "the game" is this: you make more product than you know you can sell, and then the rules of absorption accounting allow you to place the full-up cost of all of those products that you cannot sell on the balance sheet as an asset. Thus, you have fewer costs in your P&L (profit and loss) this month, and your operating income looks really good. Wonderful! There will be bonuses this month for everyone! But is the business really any better than it was last month?

Meanwhile, the financial people have been taught to manipulate (OK, perhaps "play with" would be a better choice of words) the numbers of the business in spreadsheets, and to spend all of their time booking transactions and using the "financial" accounting system to attempt to "manage" the business. The financial folks love to "allocate" costs. This month they allocate costs based on what one accountant believes is a "good" cost driver, and next month someone else in accounting believes that another cost driver would be better utilized to allocate costs. And usually the cost driver is labor dollars—even though labor is only 20 percent of the cost of the business today. All they have done is play what I call "the walnut shell game." This usually implies that whichever product has the greatest revenue gets allocated to it the majority of the costs. Now I understand why some folks today refer to a business as a "social system."

My earliest degree is in Mathematics and Statistics so I understand numbers, spreadsheets, and analytical models such as regression models (not that any CFO would ever trust a model without an "equal sign" in it). And I know that a person can always "make" whatever number he has been given as a goal. Thus, it is best not to run a business by the numbers, yet this is what we learn at our universities. In both functional groups, OPS and accounting, people are behaving as they have "learned" in academia. I know this because I teach as an adjunct at a university across town in the evening. Therefore, changing behavior, which is the purpose of my life in business, must really start in academia. Will that happen anytime soon? Probably not. And that is why, in my career, my purpose is to highlight behavior and how we change it to make business more successful and more effective. If we are going to change behavior, then our traditional standard absorption cost system (which creates all the wrong behavior) must cease for the purpose of managing the business. However, *when* one does this—that is, the timing of the removal of the standard cost system for managing the business and the implementation of the Lean cost management system—is critical.

Some in business today would focus on the numbers; others would focus on management. Still others would focus on culture and change. But it is not about numbers; it is about processes and how we achieve success. It is not about management; it is about leadership. And it is not (initially) about culture and change. Instead it is ultimately all about changing behavior. Thus, I choose to focus on behavior in order to "raise the bar" in this business. And so my story unfolds …

As I pull into the lot I see Mark, the COO for this plant—the man who has been chosen by corporate to be my partner in this Lean endeavor. Having been on the job for just a week now, I can already feel that Mark will be my best advocate and business partner in this change management initiative, or what I call a "turnaround" effort. Mark has been in his job for many years, and came here with a wealth of experiences from other successful Lean transformations. Although he is an engineer by trade, I will not hold that against him. And I smiled.

"Hi, Mark! Good morning to you! What's up?"

"I was just thinking about that town hall meeting yesterday," Mark said. "There definitely needs to be a new management paradigm instigated here in this plant from what I heard yesterday. Nancy, the CFO began and ended by telling people that she had 'reclassed' some assets, and so now we are 'looking good' for the month. You are the bean counter. What did she mean by that remark, Liz?"

"Hey! Hey! I am not a bean counter. And I am also not what some others call a 'database banger.' I have other 'faces'. What do I have to do to get you OPS people to change your stereotype of financial people?"

"For starters, you need to lighten up, Liz. I was just teasing you."

"OK, good point. Oh, by the way, do you wear that pocket protector to church?" and I winked. "Now, what is it you want to know about the CFO's remarks at the town hall meeting yesterday?"

"I want to know what she meant by 'reclassing' some assets, and how would that help us—as Jack Welch would say—to 'make-the-month'? Do we even want to do that?"

"You are correct when you say 'make-the-month,' Mark. Even you can see that she likes to just grind away at the numbers, and work in spreadsheets for short-term results. She is what you OPS guys would call a 'financial superstar'—or I can also imagine you guys referring to her function as one of 'financial reengineering.' It will take some effort to get her to change the way she thinks."

"I agree. But what did she mean by reclassing some assets, and is that even ethical? Ethics is one of our core values, is it not? At least it should be! How does this reclassing impact the financials, Liz?"

"It's like this, Mark. Most assets on a balance sheet are recorded at their original historical cost, even though the financial powers-that-be would like it better if we change that to record assets at current market value. And, because of all of this, lenders (such as our bank) have come to place a lot of emphasis on balance sheets. Some financial folks—and I emphasize the word *some*—may 'play with' (a better term than reclass) balance sheets. They may do this directly by booking incorrect accounting entries and then reclassing them next month. Or they may do this indirectly by keeping transactions off the books completely. This 'off-balance sheet' accounting usually concerns leases, but I do not believe that is what Nancy was referring to in her remarks at the town hall. I believe she was talking about reclassing via the direct method, and I assume this has to do with our bank. You do know that they manage this plant and this business based on a rather large revolving line of credit, right?"

"Yes, I have heard about that, but I know very little about it. This is why I am asking you about it. I find this all most interesting. Keep talking, Liz."

"Well, the motives for what I would rather call 'playing with' the balance sheet relate to reporting requirements established by lenders such as our bank. Our bank revolving loan covenants require that we meet certain financial ratios. If the bank determines that we have not met or have violated any of these ratios or covenants, the bank could accelerate the repayment of our loan or refuse to loan any more money to us. Thus, our CFO and CEO might wish to increase the stated value of short-term assets such as Inventory or Accounts Receivable."

"Why is she bringing this up in a town hall meeting, Liz?"

"Probably because it is very much on her mind at this time. But you are right. The town hall was not the place for this discussion."

"If this is unethical, it sounds like it might be difficult to prove," Mark commented.

"Yes, it may be. Years ago a CFO asked me to inflate inventory by a penny a pound. I resigned from the job. Personally, I prefer to be able to sleep at night. But let's take an example. Suppose the ratios that the bank requires in the loan covenants are debt-to-equity and the current ratio. The financial person who is putting together the financial statements might fail to record some liabilities, thus keeping the debt-to-equity ratio down. Or they may inflate certain assets, such as inventory, to keep the current ratio up. While there has been talk of changing how we value assets, you might write down an asset (such as equipment or obsolete inventory) but you never would write 'up' an asset—especially not inventory! The most common method I have seen to increase the current ratio, or any ratio for that matter, is to 'play with' accruals. And you are right, Mark. This would be difficult to prove."

"Liz, I really do not believe that Nancy is smart enough—or experienced enough—to know how to 'cook the books.' These little tricks must be coming from Jim. I will remember to bring up this banking stuff casually the next time we play golf."

"That would be good, Mark. We cannot have a leader with poor or unethical behavior," I said. "And Nancy should have better things to talk about in front of the entire workforce. How did we really do last month? Where are we headed next quarter? How can anyone here help improve the business? These are things I would be interested in hearing … rather than the skinny on loan covenants! And I would be more interested in seeing the company P&L rather than this calamitous balance sheet. Balance sheets are the main focus of external users of business information. For you and me and other internal associates of the company, the income statement, or P&L, would be of greater interest."

"Are you ready for the plant walk-through today, Liz?"

"I sure am, and I hope that this will be the beginning of the end of managing strictly based on the numbers, and I hope we do get some of these management folks to start to see what is really happening here."

"I am going to get organized, and then I will meet you at the entrance to the shop downstairs at seven o'clock. Remember, Liz, lighten up. The OPS people will like you more!"

This had all started my first day on the job. The HR manager, of all people, wanted to see in a spreadsheet the reason why the plant efficiency rate is only 37 percent. At the end of my first week, I concluded that I could not show this in a spreadsheet—nor did I want to. So we are meeting a few of the top management people this morning at the start of the first shift for a walk-through the manufacturing operations. While seeing an HR manager on the shop floor could make workers nervous, I can think of no better way to illustrate our poor efficiency rate.

"Hello, Helen. How are you this morning?" Helen is the HR manager. She is an older and very traditional person. Already I have perceived that Helen's core values do not include communication or diversity of people and ideas. Serving the customer and having fun doing it would also not be a core value of hers. And Mark thinks that only finance people need to lighten up.

"Good morning, Liz. Just what are we all doing here?" "We" included Helen, Mark, Tom, the plant manager, and me.

"We are just starting up the plant," said Tom. "Here is where we ship finished product. This is as close to the customer as we can get. We will start here and work our way to the beginning of the manufacturing process where they are making the first batch of parts."

"Why do we have to do this? Why can't you just give me a spreadsheet to explain the poor efficiency rate of 37 percent?" asked Helen. HR in this plant, as in most plants, does not want to deal with a manufacturing revolution. But is it not the job of HR to talk with people? Rather, they seem to prefer to detach themselves from the plant and the workers, and to look in spreadsheets for solutions to problems. I guess they are no different from so many others here.

"OK, let's walk the line," said Tom. First, we see a lady with her head down on a table. Helen tapped the woman on the shoulder, and asked her what she was doing.

"I am waiting for that first batch to come down the line," she replied.

"When do you expect that to happen?" asked Helen.

"Well," the worker said, "that first batch usually comes through about one o'clock in the afternoon, right after lunch."

We continue our walk, and come upon workers sitting outside a break room, drinking coffee. Next, we see a group of people reading the morning paper, talking about how they should really be back home "hayin'." Did I mention that this plant is situated in a small farming community? Now we are back to the top of the line, where they were making a batch of parts—which they are at the moment scrapping, and starting over!

"Now do you see why we have an efficiency rate of 37 percent?" I directed my question to no one in particular. "How many people did you see adding value to the product?"

"OK," said Helen, "but what do we do to improve this situation?" I was waiting for Tom to answer this question, but he was obviously not going to touch this issue with a forty-foot pole or a pitchfork.

"Helen, it is not just a 'numbers' problem, although that would make it easy to keep score. And it is not just about machines—although ours are very old and have never been maintained. Nor is it just about inventory, which is really hard cold cash—even though we have much too much of that stacked everywhere but in the bank. It is about people! So it is also a 'culture' problem, and we simply cannot explain that in a spreadsheet," I replied, "even though machines and people and inventory translate into cost. Efficiency, Helen, implies keeping people busy doing anything—just to cover the overhead. Efficiency means having good people do wrong things well. We should be concerned with effectiveness. Effectiveness looks at what is and what could be! Effectiveness means having people do the right thing well, and that adds value."

"I think we need to revisit our core values," said Mark. "People here do not feel empowered. The factory has been run with the philosophy of 'do your job, do what you are told, and keep quiet.' I know that is not your Lean philosophy, Tom."

"No, it is not my business philosophy, Mark. But how do we empower people? Do we just sprinkle them with pixie dust? How do we get the whole company to work together as a team?"

"Well," I answered, "it might start with our HR system where we appear to be managing people by fear. I would like to talk to you sometime, Helen, about a *systems merit* approach to people as opposed to the *pigeonhole* system we have now. Perhaps we could discuss this—just the two of us—and then, if you agree, bring this up in a meeting with all key management players."

"Good idea," said Mark. "Gotta run! Will I see you all at the board meeting later this afternoon?"

"You bet," said both Helen and Tom. "We have more problems than we can count. And our supply management and purchase price variances are just one of many problems to be addressed at this meeting."

"And I am sure we will see some great spreadsheets at this meeting!" Mark said to no one in particular, with a smile on his face.

After some time with the staff in the accounting office, I asked the CFO, Nancy, if she would like to get some lunch across the street. "Sure," she replied. Nancy was, in my opinion, in over her head. She had a BA degree which she had earned over thirty years ago, and she had never updated her education nor worked anywhere other than at this plant. And, when anyone challenges her, her first response is to cry. Perhaps none of this is her fault. Perhaps she is the by-product of the CEO, Jim, who also utilizes the same business philosophy on her as on the shop people, "Do your job, do what you are told, and keep quiet." Jim seems to truly believe that the CFO's job is an accounting job, period. He has no idea of the many "faces" of a modern-day, world-class financial manager, and he has admitted that to me in private.

I had started working on Jim the very first day on the job, suggesting to him that a financial manager is more than just a bean counter or bookkeeper, running numbers in spreadsheets and booking debits and credits. I brought this up with Nancy over lunch. I wanted to hear her response. "Nancy," I said, "I see all financial persons falling into one of three categories: first is the typical bean counter or the person who just runs with numbers in spreadsheets and books debits and credits; second is what some call the database banger or the financial person who really is talented in analyzing data to get information, and knows something about the overall business; and third is the financial person who is acting as business partner to Jim and other top management people. Into which of these three categories do you see our accounting associates fitting?"

"I suppose we fit into the first category of the numbers manager," Nancy replied. "However, we are just responding to Jim's expectations. 'But financial reports are all I need to run the business!' Jim says to me. 'I know the reason for the numbers. I do not need a bookkeeper to tell me that!'"

"OK," I said. "So Jim is a bit of a chauvinist, and needs to go to charm school." I remembered what Mark had said to me this morning, to "Lighten up, Liz." He also said that we need a leader, and not just a manager. I will let Jim's comment to Nancy pass for now as I believe actions will speak louder than words in the board meeting later this afternoon.

Nancy and I had a nice lunch, and I did not bring up the negative cash and negative inventory that she had on that disastrous balance sheet this month—the balance sheet she had shown at the town hall meeting. No one else noticed this, not even Mark. Mark even exclaimed to me after the meeting, "We have negative inventory!" as if that were a good thing and something to be proud of. I wanted to respond to him, "How can we have negative inventory when we have so much stuff piled everywhere—even in trailers all over our campus?" But I said nothing. There would be another time and place for this discussion.

"Nancy," I said. "Thank you for all of your help in getting your staff to come up with the data we needed on purchase price variances for the board meeting this afternoon. We hope to make a strong statement as to the need for OPS and the procurement people to work together."

"You're welcome," said Nancy. "I am always glad to help." Nancy was always glad to help with anything related to the office, but she would never put the suit and stiletto heels in the closet and walk into the plant. I can still remember my first day here when the plant manager, Tom, approached me and said, "Well, are you going to sit up there in your ivory tower and play with the numbers, or are you going to come down here and look at the way

I run this business?" If Tom had said that to Nancy, she would have burst into tears.

"Let's get back to the office and then head over to the boardroom for the afternoon meeting," I said. "You do not want to miss this one! From what Mark told me, this meeting should be the start of something new here."

On my way over to the board meeting, I saw Mark getting ready to cart his data (literally) over to the admin building.

"Hi, Mark. Are we all set for our presentation? I just had lunch with Nancy, and told her how much we appreciated all the data she gave us on variances in purchase prices."

"Yeah, well, I think the shop people have given us the data we really needed for this presentation."

"OK," I said. "Let's go make a statement—without a spreadsheet! Do you think they are beginning to get the point that managing by numbers is not the optimal way to run a business?"

"I hope so," said Mark. "I had one of the guys bring in a wheelbarrow from his farm. As they say at Budweiser, 'here we go!'"

The board meeting was attended by Jim, Nancy, Mark, Tom, Greg, our Lean leader, Bart, our inventory manager, Susan, our head of procurement, and Helen. Nancy put up her spreadsheet on variances to standard, and highlighted all of the unfavorable purchase price variances. She is not to blame here for lack of information (as opposed to just pure data). After all, she is just doing her job as her boss, Jim, has dictated.

After the spreadsheet had been posted on the wall in high definition, in strolls Mark with the wheelbarrow containing hundreds of work gloves! Everyone in the room is looking at this with horror, and wondering what in the world Mark is up to with this very used piece of farming equipment. Every factory glove in the wheelbarrow had a tag with the name of the business unit that bought and used the glove, the supplier's name, the number of units ordered, and the price. Mark went a step further and dumped the contents of this wheelbarrow onto the boardroom table. As management team members began to pick up and examine these work gloves and their tags, they noticed that different business units were purchasing the same glove from the same supplier at different prices. And there were many other anomalies. Then Mark brought it all together by saying, "I think we have far more than variance analysis work to do to expose what is going on here." Yes, there was a story behind the numbers, but not a story that Jim or Nancy had been aware of, and not a story that was visible in a spreadsheet.

Back in Mark's office, he and Greg, our Lean leader, and I discussed the next step in our plan to change the way "the work" works, how people see the business, and the behavior of our associates in this business. Today had been one full day and tomorrow promised to bring more insight into the "new management paradigm" as Mark had referred to this.

"Mark and Greg, how do you think that went? Are they getting the picture? I think the next step in opening their eyes is to take them down into the staging area, where we stage all of the many cartons and boxes for packing products to ship to customers. What do you think?"

"Good idea," said Greg. "They will see that we push huge quantities of every size box and carton variety we have into that small staging area. We have cartons and boxes piled wall to wall, and floor to ceiling. Then the forklift guy comes in to get whatever they need for the day, and pokes holes in every carton or box in the room! What is that doing to our scrap rate?"

"What scrap rate?" I said. "Did you see all of those parts this morning that were not made to spec? Well, they just dump them in a pile in the yard, and wait for someone to haul them away. We have no financial tracking system for scrap."

"I did not know that," said Mark.

"You would know that if our offices were adjacent to one another on the shop floor," I said. "And procurement would know what is really going on if they sat with the OPS people in the plant. We really need to get started with a new 'seating chart' around here! It just might improve our efficiency rate, our scrap rate, our morale, and a lot of our other behavior problems."

"Let's talk more about that tomorrow. How would that 'seating chart' help us?" asked Mark and Greg in unison. "Is that anything like an org chart?"

"We'll do that," I replied. "But, before we discuss that seating chart, we need to take a visual of that staging area first thing tomorrow morning."

"Don't you first have some numbers to crunch, Liz?" said Mark with a smile.

"I will do that tonight, Mark, right after I think more about how to become more proficient in banging databases and how to get these accounting folks out of their glass offices and into the plant. They should not be just analyzing costs but seeing the reasons for all of the costs, including what I would call soft costs such as lead time, delivery time, cycle time, etc. Did I ever tell you about the large company that ordered all products for its retail business from China where material was cheap? The controller of that business was in one of my evening MBA classes at the university and told the class that, because the material was picked up and put down 31 times between the time it left China and the time it arrived in the States, the price was higher than quoted and the material arrived behind schedule."

"Yeah, I did hear about that, Liz. And I also heard that this company had to file for bankruptcy."

"Yikes! I do not want that to be us! See you tomorrow morning, bright and early, at the staging area. Have a good restful night, Mark, and thanks for everything today."

The next morning, at exactly seven o'clock, I met Mark at the staging area. This is a relatively small room, and it had been filled floor to ceiling and wall to wall with boxes and cartons of every size that we used to package any variety of finished products for shipping to the customer. At the start of the first shift, the room was totally filled, and the forklift driver was on his way in to get the cartons and boxes needed for the first eight-hour shift. There was just one small problem. No one knew what was to be packaged and shipped today!

Harry, the forklift truck driver, was hollering at anyone who would listen, "Hey, does anyone know what we are packaging and shipping today? What cartons or boxes do you guys need?"

"I don't know, Harry. Perhaps you should ask Tom. I hear that someone either faxes or emails him a packaging/shipping plan at the start of each day," said one of the workers.

"Good grief," I thought out loud. Here are all of these people ready to package, box, and ship product, and no one knows the game plan! "Mark, how can this be?"

"Just add this to the number of things we are learning about this company," said Mark.

"OK, but what do we do now?"

"I think we need to find who has this packaging/shipping schedule, and make it public ASAP so that everyone knows what the game plan is for the start of today."

"Then what?" I asked.

"Then we quickly put up what I call *production control boards* for the folks. This will tell each operator what he needs to package each hour—the target. Then the operator posts, next to each hour, how many cartons he actually filled, and there is even a space for the operator to tell why he filled more or less than the target. This eliminates the need for me to read labor reports at two in the morning. I can just walk through the area, and see what is going right or what is going wrong, and make corrections right there and then."

"Sounds great, but just one comment," I said. "The very first week I was here our corporate HQ had sent Mr. Sato, the corporate Lean sensei consultant, to this plant. At the end of the week, I saw them carry him out on a stretcher! The word on the floor was that he had had a stroke."

"I know," said Mark. "Some at corporate did say it was a stroke. I am about to have a stroke here myself!"

"But did you see what the factory workers did to the electronic production control board that Mr. Sato put together for us? They completely tore it down! You should have done something, Mark!"

"What could I do? This place is run like a prison, and the prison guards are the CEO, CFO, and the HR manager. They brought that consultant in only because HQ demanded it. Then our prison guards will tell HQ that this initiative by the Lean sensei was a failure because we don't need any Japanese guy telling us what to do. Remember that it was you, Liz, who said we do not have diversity as a core value here."

"We really should start engaging these top management people, Mark," I pleaded. "I've spoken with some folks at an auto manufacturing plant in Europe who tried to implement Lean bottom up, and it didn't work. And I do not think our middle up is going to work either. We need to get top management—those you refer to as the prison guards—to buy in, and sign off that they are fully committed to this Lean change management initiative."

"I have tried talking to the CFO, but every time I do she cries!" Mark exclaimed. "She blames me for everything that is wrong. When I press her about going into the plant, she runs to the CEO and tells him that I won't let her in! Jim always takes her side, and please believe me when I say that he is not 'Mr. Suave' with everyone."

"I know, Mark. Did you hear Jim at that OPS meeting last week? Our best product line manager, Mindy, was to pitch her team's performance results. She is timid, but I hear she is smart, and she really knows the business."

"Yes, she is smart, and she does know the business," Mark commented. "When I first came here she was sitting in a corner in the break room, playing cards on her laptop. Once we got her engaged, she was dynamite! She does not have a college degree, but she is passionate and soaks up anything you give her like a sponge! Being passionate is a desirable quality of a good leader."

"Well, when she sort of froze up at the start of her presentation, Jim says, 'Hey, did you hear that female chicks are slower to peck their way out of the shell than male chicks? I read that in a book to my 6-year-old just last night!' This guy gives a whole new meaning to the word chauvinist. No wonder we have morale problems!" Liz exclaimed. "And then, too, our own plant Lean sensei, Greg, is not helping much either … especially in the office with the accounting staff. He has initiated a book club, to read and study about Lean. How will that help? That is worse than mandated training courses."

"I'm working on Greg," Mark explained. "We will soon be starting Learning Kaizens, where every person working here must participate in a weekly kaizen every six months. Each Learning Kaizen will be two days of Lean training, followed by three days of Lean implementation. And no more book clubs! That should help."

"OK, back to the present crisis," I said. "Would you agree that our only recourse right now is to put up these paper production control boards? To be different, our boards could be called Takt boards—that's a German thing! And they are manual as opposed to the electronic boards Mr. Sato put in place. We will stay with the operators through the shift, and make certain they understand that this is helping them!"

"OK, I agree," said Mark. "While I get the Takt boards in place, why don't you engage the supplier of those boxes? He is here on our campus. Get him in here, make sure he gets the packaging/shipping plan every morning before the first shift starts, and encourage him to put in the staging area only what we need for the first few hours, to avoid the forklift driver punching holes in everything in that room."

"Will do," I said. And at about that same time I saw product all over the floor at the back of the line, and heard a worker screaming for help! Finished packaged product was coming off the line, and Paul is supposed to be taking these packaged cartons as they come off the line and place the cartons in large boxes for shipping to the customer. Every time he tries to move a full box to a pallet, the bottom of the box falls out! Now he is far behind the plan for the shift, and products are just crashing off the line and all over the floor. It's piling up out there!

"Someone go help Paul!" I screamed. "Then someone get the box supplier in here!" Not only do we have purchase price variance problems, we have supplier quality problems as well. I feel like a gopher on a golf course.

If I can just move about while keeping my head covered, I may be able to avoid getting whacked again today. I am thinking that quality needs to be yet another one of our core values—along with valuing people, valuing diversity, valuing teamwork, and valuing communication. Of course, empowerment comes into play here also. Then, too, I am thinking, "How many 'faces' have I worn since I began this job as controller?"

That night, at home, I started thinking more about empowerment. People are empowered when they do not feel threatened, and when they do not feel fear. What creates fear? And how do we dispel fear? The CEO and CFO should be saying to people, "Does it make sense to you? If so, let's do it. If not, let's talk about it." It is "how" we resolve conflicts—the design and process of problem solving—and not "who" resolves a conflict that counts. Perhaps this is the discussion that Mark and I need to have with the OPS people and the supplier, both of whom were the triggers of another very busy and stressful day.

The next day Mark and I compared notes from the day we spent dealing with problems in packaging and shipping, with Takt boards, and with suppliers. I tell Mark, "We need to focus more on empowerment, and the reasons for fear in this plant. And no, I do not mean we sprinkle everyone with pixie dust. Empowerment implies identifying who owns what and who has responsibility for what processes."

"I agree with you," said Mark. "Empowerment and teams can actually increase controls, Liz, something you bean counters are always fussing over."

"Do not get me started, Mark. I am not a control freak. I would agree that we financial types need to become less like policemen. We need to learn to be coaches and business partners to other decision makers like yourself, as well as agents for cultural change. We financial types need to focus on establishing a 'value' attitude throughout the business. I would like to think of myself, not as a cop, but as a catalyst. That is the 'face' that a financial manager needs to wear every day!"

"OK," Mark agreed. "So we have empowered factory workers, as of yesterday, to speak up when they need information, such as the packaging/shipping schedule. And we have empowered workers on the line to ask for help when things go wrong, such as boxes falling apart due to poor quality or due to our procurement team setting themselves up for favorable PPV (purchase price variances) without regard for their actions. And we have empowered our supplier to be responsible for the processes of filling the staging room with only the cartons and boxes we need in the immediate future as well as only quality boxes."

"Why do you think these people never felt that they could speak up before yesterday?" I asked Mark.

"From what I have witnessed over the years, the answer to that question is fear. Our accounting staff fears Nancy, and everyone fears Jim and Helen," Mark replied. "Proof of that is in the high turnover rate here. Nancy and Jim have never heard of empowerment, or simply do not believe in it."

"Do they understand that empowerment does come with boundaries?" I asked Mark.

"Years ago I heard of an IT manager who was empowered to reprogram the process of placing customer orders. He wrote a program that essentially said to take everything out of the warehouse except the order, and ship this to the customer. Needless to say, both the customer and the warehouse replenishment team were shocked by the size of that one order!"

"That's a great example of empowerment-gone-wild!" I exclaimed. "We have enough wild things happening here, Mark. We need to help management understand that Lean is a long-term inclusive initiative and that we do not need managers but leaders, and we need leaders who are not dictators. Plus, we need to see leadership coming from the top down."

"So what are you thinking?" asked Mark.

"I think we do not want to focus on culture as a target in our Lean initiatives right now, but instead first focus on behavior. You and I both know, Mark, that if one focuses on long-term goals, the short-term goals will come automatically. Likewise, if we focus on behavior, a change in culture will come about automatically. And a Lean accounting system, along with new Lean metrics, does drive behavior. Also, an HR model such as systems of merit can be helpful in changing behavior by not making people feel that they are pigeonholed in any one job—that they are free to engage in learning within the organization to get them to any level or position in the business. I really do need to have this discussion with Helen soon."

"Good! And also we need to have dinner some evening so you can tell me more about this Lean accounting. Metrics I understand and I agree we need new metrics if we want to gauge our success in our Lean endeavors. But Lean accounting—or any accounting—is beyond my area of understanding. All I know about accounting I learned in my MBA program."

"That could be good or bad, Mark, but perhaps we first need to talk about how we restructure this company. Right now we are structured in functional silos, and this is not encouraging the correct behavior to help us with our Lean change management. In our 'silo system' everyone is a specialist, and no one can see the big picture. Each associate is optimizing just his or her one small piece of the business. Right now we all are thinking about point improvements, not systems. We are so myopic! We don't think about how to organize and prioritize the pieces of the entire business from the perspective of the customer. Would you agree?" I asked Mark.

"Yes, I believe you are talking about tearing down functional silos rather than paving the cow path between silos, and then establishing what I call product families. This would also reduce fear as people would be more engaged in learning in their family rather than engaged in compliance in the business as a whole. However, next week we need to focus on getting top management engaged in leadership and getting all of them on board as a team. We cannot drive this initiative, Liz, from the bottom up or the middle up."

At this point, I am glad that I have Mark as a partner in this challenge— even if he does have a "country" way of expressing himself. As we would be working together for one or two years, our corporate HQ had arranged a meet-and-greet for us before I accepted the job. Mark had insisted that we meet casually at his ranch which is located just a few miles from the plant. I liked Mark the minute I met him! He has a positive attitude, he is upbeat, he's high energy, and he is always wearing a big smile. When I complimented him on these attributes and asked how he stays so positive, he replied, "I try not to say 'no' too often!" and he said that with a big grin on his face.

In his spare time at home on his ranch, Mark works with dogs (and their owners) with behavioral problems, and he trains dogs. Whether your dog attacks everyone who comes to your door or your dog buries his bones in your neighbor's rose garden, Mark can train the dog to lose the bad behavior and retain the correct behavior. He coached me, while I was there, on giving instruction to one of his dogs. I must admit (but had not told Mark) that I have had a fear of dogs since I was attacked by a Doberman as a small child.

"Don't pucker your brow when the dog is executing the instruction you have given to him," Mark said sternly to me.

"I'm not puckering!" I said.

"Then what is that 'face' you are making, Liz? Surely that is not your smiley face? You are confusing the dog with that angry face. You cannot show fear or anger when training dogs."

I am thinking that this must also apply to people in business. And it is obvious to me that Mark's skill in dog training spills over onto people.

But this is just one of two reasons why Mark is so good with people, and changing behavior in people—or in dogs.

While discussing our backgrounds at our first meeting, Mark told me that he grew up in a blue-collar family. His dad was a machinist and head of the union at his shop. As dad was union leader, Mark told me that there were frequent union meetings in his family's living room. It was obvious to me that Mark had been observing behavior—in people and in dogs—since he was 8 years old. Thus in Mark's family, as well as in the product families he creates in business, white collar is on the same footing with blue collar. There is no distinction between the office worker and the shop floor worker. Yes, Mark is what I would call a behavioral specialist, and I am glad to be partnering with him in this Lean endeavor.

2

Systems of Merit

I should have studied to become a psychologist! Our present HR system does not fit with Lean. Our current HR system manages people with fear and carrots and sticks, but Lean is all about developing people. A Lean HR systems of merit will better help our associates achieve their potential by emphasizing that we are a learning organization.

The first few weeks had been nothing but putting out fires. Firefighting will never go away until we stop executing "point improvements" and create a "business system." As I mentioned this to Mark this morning, he asked, "Did you remember to bring the truck today?"

"Very funny," I said. "Seriously, Mark, we need to back up the truck, and think about company values and engaging top management in leadership rather than just management. This business does not need to be 'improved.' It needs to be 'reinvented!'"

"OK, so what do you mean by *values*? And how are values different from *principles*?" Mark asked.

"Principles are things your mom taught you, like honesty, integrity, and respect for others. Values reflect an individual's background—or, in this case, the background of a business. They are shaped by our experiences. Another way to look at this is to say that principles are external, while values are internal. Values will guide us to where we want to go, and so we must all have the same values, and those values must be the values of our customers as well."

"We do not want to get into a war of words with shop people, Liz. They will have no patience for this egghead type of conversation."

"This is very important, Mark! You could not even run a small neighborhood grocery store today unless you first had some core values. Let's say your core value is to be the most productive grocery in your neighborhood. Then you will need a strategy to reach that value. If you just open a grocery and start selling stuff, you will probably fail! And this is without a doubt one big reason why we are failing. This idea of creating values first is not just a lot of fluffy words, Mark. This is important. And our values need to be decided upon by top management, and then communicated to everyone in the business. One needs to live one's values!"

"OK, I get it, I think," said Mark. "We need a cultural change, and by adopting core values we will be instilling this in our people as a prerequisite for implementing Lean."

"That is correct, Mark. However, we do not necessarily have to *instill* this in people. We may just ask those associates who do not believe in our core values or change to leave the organization. They will not be happy here. Then we hire people who we know embrace our core values."

"I have heard of the term MBV which has recently replaced MBO in Lean organizations. Is this 'management by values' the same thing you are talking about here?"

"Yes, it is, Mark. MBO, or managing by objectives, is a process of defining objectives within an organization such that management and employees (but not the customer) agree to the objectives and understand what they are. MBO is essentially each person optimizing his or her small piece of the business. But in Lean we want to optimize the system! So MBO encourages myopic behavior. Also, MBO usually gives us 'number' targets, and any OPS guy who has earned an MBA can usually come up with a way to 'make the number.' You know that!"

"I get it," said Mark. "So MBV, managing by values, is more concerned with goals where 'the process' matters, and where we are more concerned with the approach to achieving the goal—or, in this case, the core value. And I assume that the core value is not a number, am I right?"

"Yes, you are, Mark. It is just like with metrics! For example, reducing scrap by 50 percent is better than saying we want to reduce scrap by $100,000. We want something that is actionable and cannot be manipulated by anyone. Think of the five Principles of Lean: (1) create value for the customer, (2) map the value stream, (3) flow, (4) pull, and (5) perfect! It all starts with value!"

"And value begins with the customer!" Mark exclaimed. "When I say that quality should be a core value for the company, I mean that the customer requires a quality product or service."

"Yes, and that is yet another reason to not utilize MBO in Lean organizations as it places more emphasis on the setting of the goals rather than the plan to get there. What will we use—what behavior will we instill—to get people to meet our quality value? We want to avoid employees 'gaming' the system and thus distorting results."

"So empowerment does come into play in MBV. Rather than setting objectives from a long list of choices that management has given us, we are setting values given to us by the customer, and then asking the employees what the plan should be to get there," Mark commented.

"Yes, Mark, and even a friend of yours, Deming, encouraged managers to abandon managing objectives in favor of leadership. This is just one more reason we need a good leader and not just a manager. For that matter, we all need to lead by example. We do this with our kids at home. I tell my children they must get good grades, and then I set the example by having a place and time where the children and I study for our classes together. I messed up an auditing course not too long ago, and received a C. My excuse for the poor grade was that I saw no purpose in auditing, but I could not say that to the children. When I got home that night, my son had taken the small television out of my room and left a note that said, 'When your grades improve, your TV will be returned.'"

"That's funny, Liz. OK, you have sold me on MBV rather than MBO. But now can you give me some examples of values?" Mark asked.

"Sure! Values are things like teamwork, communication, and diversity. For us, delivery is a core value. Our customers know we can deliver on time—even before our competitors. Quality, as you have mentioned many times now, might also be a core value, as well as technology. A company value might also be something like 'make money and have fun' which I imagine just might be a core value at Tupperware, or 'concentrate on cash flow, not profits,' or 'preserve an atmosphere of a young entrepreneurial company' which is probably the core value at a company like Google. Another example of a core value might be 'pride in product' which I imagine might be a core value at almost any company."

"I believe a good core Lean value here might be 'no single right answer,'" said Mark.

"There you go! You are right on!" I replied.

"So we need to have this conversation with top management, and decide among us what our core values will be. What do we feel the customer wants from us? This is what will drive our business going forward. I know our customers well, Liz. I meet with them frequently. And I know the CEO very well. Jim really is a good man—not a chauvinist, and not someone who makes fun of others—and definitely not someone who insists on cooking the books. He really believes in Lean change management as a way out of our troubles, and he will stand firm in that decision. He just does not know how to get this started. He believes we can and should start with operational improvements in the plant. But he is aware of behavioral problems here, too. I will talk more with him about how we get started with our Lean initiatives."

"Uh-huh," I said. "And Jim has told me that Nancy has respect for me and where I have been in my career. While you are talking to Jim, I will talk with Nancy. I do still believe that she is in over her head. And she has no soft skills. I have been told by Helen that there is 100 percent turnover in her department every year! What does that tell you?"

"It tells me that people either hate her or fear her!" Mark said. "I would not want to try to talk to her. Well, I have tried, and she gets upset and cries. I told you about this. Then she runs to Jim, and he protects her. Do you think they have a thing going on?"

"Oh, stop! You engineering guys have a vivid imagination. I believe they are very close because they have both been here since the beginning of their careers—over 30 years! They are accustomed to one another, and each knows how the other reacts in any situation. If I were in charge, I would replace her with someone who has updated her education recently at one of the few universities that teach Lean, someone who is certified, and has some soft skills.

"The problem seems to be," said Mark, "that creating 'value' is something new for both a business and a university. 'If it ain't broken, don't fix it!' Businesses and colleges don't seem to know that what they have is broken."

"I know, Mark, but I am not in charge of employee relations, here or at the university, so I will just talk with her while you talk with Jim. Then we bring all top management together, decide on our core values, and set the stage for a company Steering Team managed by core values and leaders, embarking on Lean change management endeavors, to turn around this company."

Later that day, Mark talked with Jim, and I had individual conversations with both Nancy and Helen. In my meeting with Helen, I brought up the concept of an HR model I call "systems of merit," which has been used successfully in Lean companies. This will be important if we truly want to get people out of the "pigeonholes" they feel they are in, and should help to motivate and empower the workforce by eliminating fear of failure, fear of making mistakes, and fear of competition between employees. I wonder if Helen is old enough to remember the TV episode "Lucy in the Chocolate Factory," I thought to myself and smiled.

"Hello, Helen. Thanks so much for taking the time to talk with me regarding the systems approach to HR which I think might help with motivation and empowerment here. I truly appreciate the opportunity to talk about this with you," I said.

"You are welcome, Liz. You did have a good point regarding the reason for the poor efficiency rate in the plant. I am eager to hear about this systems approach to HR."

"Systems of merit, Helen, was developed by a business engaged in Lean change management when the traditional Myers-Briggs model was not only not helping but hindering the company's Lean transition. I am assuming that we use something like Myers-Briggs here?" I asked.

"Yes, we do," Helen replied. "We test every person coming into the business. The Myers-Briggs model tells us, for example, whether a person is an introvert or an extrovert, and then we can place them into positions that best fit their personality." I am already thinking of what I learned from Mark and his dogs. "Behavior trumps personality!" Mark had told me. I am wondering if they also test for "tunnel vision" when an engineering new hire goes through medical. I will have to ask my good friend Mark-san about that. That should crank him up!

"I agree that having a good personality fit will make a person happier in his or her job, but are you not really stereotyping people with Myers-Briggs, Helen? By that I mean that you must place introverts in financial and other office jobs or on a robotic line, and then you must place extroverts in sales positions, am I correct?"

"Yes," replied, Helen, "that is basically how it works. An engineer, for example, might test as an extrovert or an introvert, in which case we would decide whether to place him in a position with suppliers or customers, or just place him in product development or on the shop floor."

"Uh-huh. Well, allow me to explain the systems of merit methodology to you. I first want to preface this by acknowledging that this was developed from, and based on, the Japanese culture. You may have seen some of this culture demonstrated recently when the 9.0 earthquakes and tsunami hit Japan. The Japanese have a unique culture. You did not see them fighting over food and looting stores after the earthquake. You saw them sharing and helping one another. And they do this in business as well. This is very different from the competitive and combative relationships between workers that we seem to have here in U.S. businesses."

"I agree that our U.S. culture is quite different from the Japanese culture. So how does this relate to HR?" Helen asked.

"For starters," I replied, "the Japanese see life as a series of plateaus on the side of a mountain, and people reside on the various plateaus at various times in their lives. The very first plateau is birth, of course, and then the second plateau is what I would describe as the assembly line worker. What is important here is to know that the person at this level wants nothing more than a paycheck to take home at the end of the week. He is not motivated to go any farther up the mountain. He is perfectly content here on this low plateau. He wants to, and will, do a good job in whatever work you give him. Just know that he wants no more. An example of a person on this plateau might be "Lucy in the Chocolate Factory." Do you remember that skit from TV years ago? Lucy and her friend were to fill boxes with chocolates as they came off a production line. When they could not keep up, they simply stuffed some of the chocolates in their mouths or down their blouses."

"Yes, I do remember that, and it was so funny!" Helen replied. "Good example! Lucy wanted no more than to finish the job each day, get paid, and go home to her family."

"Well, then," I said, "the next plateau is what I would call the 'combatant.' This is the person who sees everything not to his liking, and will pick fights—both verbal and physical. This guy or gal might be compared to the speeding, lane-weaving trucker who is riding your back bumper as you both travel down the highway. You will want to slow down, pull over, and allow him to pass you and be on his way. In business, we would want to either work with this person to change his behavior or get him out of the business. With his current life experiences and values, he will always be disturbing business."

"Do you mind if I take notes?" Helen asked. "I want to make sure I understand each plateau."

"Here," I said, "is a visual that I brought for you. I guess you know by now that I am a big proponent of visuals. This is just another one of my 'faces' about which Mark continuously teases me. Let me show you this diagram, which depicts the Systems of Merit HR model" (Figure 2.1).

"Thank you very much, Liz. This sketch helps me a lot. So now we are at the fourth plateau, which you have named the 'army sergeant.'"

"Yes, the army sergeant is the level on the mountainside on which lives the guy or gal who believes in command and control. Now you know why I have named this plateau the army sergeant. This might also be the higher top management person who, at this stage of his career, believes in 'Do your job, do what you are told, and keep quiet.'" I did not elaborate any more on this person. Helen may or may not make the connection that this is the plateau on which our CEO is residing at the moment. "Then, too, this Level 4 might be the traditional accountant who feels he must have control over everything and everyone."

"What is this next manipulator level, Liz? What do you mean by this? These are strange titles for these various plateaus."

"The names for the seven plateaus were written in Japanese, Helen, and I have just tried to give each plateau an American rendition that any

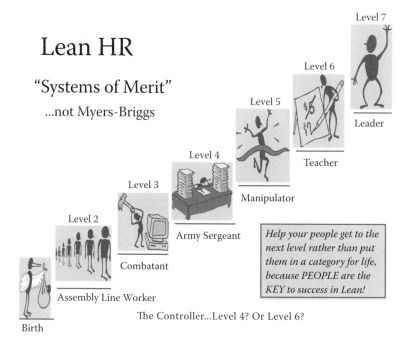

FIGURE 2.1
Lean "Systems of Merit" HR model.

worker here in the United States would understand and could relate to," I replied.

"Uh-huh. I see. Good idea. So what do you mean by the manipulator level in one's career."

"Keep in mind, Helen, that these plateaus represent not just levels in our professional life but levels in our personal life as well. It is our experiences in life in general that really determine where we stand on the mountainside of life, what values we hold, and on what plateau we stand in our business life as well. You might want to think of this systems of merit as a system similar to what the Boy Scouts or Girl Scouts have. I loved the Girl Scouts, and I know those badges were great motivators for me as a child. Whether one is a Girl Scout, Boy Scout, or Eagle Scout, there is learning that occurs at each level, the scout is rewarded with a symbol of his success in learning (a badge), and he takes what he has learned to the next level. This ultimately forms what behavior he exhibits as an adult … the definitive plateau."

"OK, I understand. So how would you describe, overall, this person who is on this fifth plateau?" Helen asked.

"The person on Level 5," I said, "is totally immersed in himself. This is the plateau where we find the majority of workers in U.S. businesses. The person on this plateau is not concerned with the business or the company in any way. This person is placing his career first and the company second. This person will do anything—and I do mean anything—to get ahead, and at the expense of everyone else in the business. This person controls his own advantage indirectly and very artfully. He or she knows how to 'game' the system. He or she will not take constructive criticism well. Just like the person at Level 3, the combatant, you want to work with this person at Level 5, the manipulator, to get him to the next level quickly."

"Uh-huh," said Helen, "and the next to the top level is what?"

"This plateau, what I call the teacher, is easy to understand, Helen. The name I gave to this level is a direct translation from the Japanese language. The person at this level, near the top of the mountain climb, is the person who understands the company and the business so well that he or she can teach this to others. We might want to consider a course for new hires that we might call 'Our Business 101' and it would be taught by an associate at this level. This person is the guy or gal who might also work with the combatant or manipulator to get them to the next level. This person has learned to be a negotiator, a business advocate, and even a catalyst for change! Hiring people who have been teachers in a prior career might be a good thing for us to consider, Helen. They would probably fit right into this Level 6 once they got to know our business."

"That is a good thought, Liz. And so what is this ultimate top level of accomplishment? I see you have referred to it as the leader."

"Yes," I replied, "this is the ultimate goal. This is the top level to which many in business aspire. This is the person who can inspire and lead people—not just manage people. But is the CEO really more influential to the values and changing culture than, say, the teacher?"

"Does one have to aspire to the top level, and then retire there?" Helen asked.

"Good question, Helen, and the answer is no. A person may reach the highest plateau, and then decide that he or she really liked being a teacher better than leading the entire organization, and go back to that level."

"This is most interesting, Liz. I see you have a note here regarding the accountant."

"Yes, I do, Helen. I would ask if the accountant should be at the fourth level, the level of the army sergeant, telling people what to do and when, or if the accountant really should be at a higher level, perhaps at Level 6, the teacher. If you were hiring a new financial person, at which level would you want that person—Level 4 or Level 6?"

"I would definitely want the financial person at Level 6 if we are to be successful in this Lean change management endeavor, Liz."

"I would agree, Helen. So what do you think? Does it make sense to you? If so, let's do it. If not, let's talk about it."

"I think that this Level 6 person is the business partner accountant that you discussed with Nancy. She told me about the three types of accountants: the bean counter, the data technology expert, and the business partner. This entire systems of merit model of HR does make sense for the approach we are taking to transform this business. I will discuss this with Jim."

"I am delighted that you see how much this systems of merit approach to people differs from the Myers-Briggs approach. And the two approaches to HR are different in another aspect, Helen. The plateaus on our American mountain climb are all pay grades. In U.S. businesses, the next step in our career is always marked by a number and a dollar sign. On the systems of merit mountainside, the plateaus are marks of advancement in our learning. So the Myers-Briggs and systems of merit models really are quite different! With the systems of merit HR model, we will reward (and our associates will strive for) levels of learning, and not levels of pay grades. If one looks up *merit* in the dictionary, the definition is 'a level of superiority and high caliber (which is) earned.' Thus *merit* implies *worth*. Then, if one looks up *system* in the dictionary, the definition is 'an organized array of individual elements and parts working as a unit.' This system of parts would also imply a system of people! So the proposed Lean HR systems of merit really is a system of high caliber people who earn their worth through learning and working as one whole body or business entity."

"I agree," said Helen. "Striving for levels of pay grades is a military thing—more 'command and control.' I don't believe we want that in our Lean business. I am curious, Liz. Where did you learn all of this?"

"Another large corporation had placed me in an extensive management training program years ago. When I looked at the schedule for the week, I saw that one whole day had been devoted to Myers-Briggs and HR. Pardon me for saying this, Helen, but I immediately went to a local pharmacy and bought a large package of No-Doze. I was expecting a very boring day. However, the man who facilitated that class that day was dynamic! He was so highenergy and vibrant, and he had the entire class on their feet all day, interacting with one another. He did everything but dance on the tabletops—or maybe he did that, too! And we participants never sat down! After class, I went up to this man to thank him for the best interactive learning session I had ever participated in. Then, out of curiosity, I asked him where he fell on the Myers-Briggs scale. He informed me that he scored as an extreme introvert. 'How can this be?' I asked him. He then informed me that anyone can get himself up for any challenge. So this facilitator tells me that he will go back to his hotel room after class, probably order room service for dinner, watch some C-SPAN or the Discovery Channel, maybe trim his toenails, and then the following morning he will 'psych' himself for another session as facilitator of this class. Mark also taught me this about interacting with his dogs. 'Behavior trumps personality!' Mark always says."

"I hope you will be joining us when we get together with Jim and Mark and Nancy to discuss core values. This is another very important 'must do' before we truly jump-start this Lean change management," I said to Helen.

"I am looking forward to the meeting, Liz."

As I am heading back to my office, my cell phone rings. It is Mark. "Hey! What's up?" I ask. "Have you spoken with Jim?"

"Yes, I have, and he is very passionate about this Lean change management as the solution to our problems. He really wants to do this, and says that he will not allow anyone to sabotage our efforts. Of course, he still talks about 'cutting costs' rather than 'eliminating waste.' But this happens with any leader in any Lean transformation whenever the economy goes sour. I will keep working on him. You had four kids, Liz. You surely understand that one must say something about six times before someone gets it. That is Jim! Have you spoken with Nancy or Helen?"

"Both," I replied.

"And how did those discussions go?" Mark asked cautiously.

"Well, the discussion with Nancy went as expected. She volunteered to tell me about the book she has been reading on Lean accounting. This is true executive arrogance, to think that reading a book makes one an expert on a subject. I believe Sato-san refers to such people as 'concrete heads.' Then she showed me the 'Lean P&L' she has developed by 'mapping' every single traditional general ledger account to a Lean P&L general ledger format."

"Yikes! Does that mean I will still see my lunch expense in every Lean P&L? I don't really need to see that, Liz."

"I know, and you will not see it. She needs an attitude adjustment, but we are not even there yet, Mark. I tried to get her to warm up to the idea of going into the plant for a Learning Kaizen, to get to know the business better, but she changed the subject and started talking about some new system for accounts payable. I tried to talk with her about 'Lean accounting' versus 'accounting for Lean.' We want to work 'accounting for Lean' first, and by that I mean we need to support the Lean manufacturing efforts in the plant and make that our first priority. Then she can 'Lean' or take waste out of her accounting functions. We also talked about soft skills and the three types of accountants. Like you said, six times may be required— maybe more! I did not bring up the extreme turnover rate in her department. And I did not bring up the negative inventory on the balance sheet. I did bring up the topic of new metrics, and she is all for that. However, if employee turnover is a metric, she wants to know when this would start?" Mark and I both laughed.

"What about Helen? Did you have the discussion with her on this new HR system you spoke of?" Mark asked.

"Yes, I did. And it went very well. She liked it! And furthermore, she is going to discuss it with Jim tomorrow. It sounds like we have a new systems of merit HR model in hiring and motivating people in our new Lean enterprise. No more managing by fear and pay grades. From now on we will be developing people's skills so that they can problem solve, get to know the business really well, and thus achieve any position they would like in the business."

"Great! Jim is setting up a meeting for first thing tomorrow morning, offsite some place, to discuss our core values and our strategy to get to where we want to go. Have a great night, Liz. Would you like me to pick you up in the morning for the trip to the offsite meeting?"

"That would be great, Mark. Then perhaps we could have lunch afterwards to discuss the actual kickoff for what you call our new management paradigm. I must also tell you that I have seen many applications from your dog training practices that would apply to people in this business, Mark."

"Thanks, Liz. I would agree that a little cheerfulness and self-assurance bring about success in training people as well as dogs. People can be trained—just like dogs. Well, perhaps I did not articulate my comment very diplomatically. I should have said that businesspeople can learn good or bad behavior. It just depends on the methods we use—both in training and in rewarding the person."

3

Steering Team Offsite Meeting

Ultimately we want to change the culture so that we place more emphasis on long-term accomplishments and focus less on short-term results. Thus, our strategy should not be about "making-the-month," as Jack Welch referred to it, but rather about improving the business over the long run. We will pull people together and lead them, not just manage them, by creating a vision of what we believe in through visuals, emotions, and—most important of all—core values. Creating value is the first principle of Lean, and one of three prerequisites for a Lean cost management system that will change the behavior of people and thus our culture.

"Thanks for picking me up this morning, Mark. This gives us a chance to talk before the meeting with Jim, Nancy, and Helen and, I hope, Tom as well. Tom wants to do the right thing, but he too is caught up in the 'do your job, do what you are told, and keep quiet' business philosophy of the 'prison guards.' I think we need to stop referring to top management as 'prison guards.'" We both laughed.

"So what will we discuss this morning? Or perhaps I should rephrase that question, and ask how do we want to lead this discussion on core values?" Mark inquired.

"I believe we want to make the point that this company's values will drive the business from this day forward. Choose the core values carefully. They must originate from the customer, and we all must be in full agreement. It all goes back to the discussion I had yesterday with Helen on efficiency versus effectiveness. Efficiency implies standing inside the business, looking out, while effectiveness implies standing outside the walls of the business, looking in, from the perspective of the customer. We need to emphasize this, Mark."

"Yes," Mark said. "We also need to emphasize that we are moving from MBO to MBV and from management in general to leadership. How will we describe this new leadership paradigm? And how will this affect our associates' behavior? What new behavior do we want to encourage in order to drive the creation of our new Lean culture? And I would imagine that you, Liz, have some thoughts on how the financial function will change and impact all of this—especially behavior."

"Yes, I do see the financial function changing dramatically, as well as the financial people becoming change agents and catalysts for change. They need to put on new 'faces.' This Lean cost management system is what will cement all that we accomplish in Lean, and ensure that our progress does not slip backward. But more important this morning is to identify the concept of a model for organizational change—much like the model of change for HR that I presented to Helen. But this time, I have no visual to show anyone. Or … just thinking about it, maybe I do have a visual!"

"You, Liz, caught without a visual? I cannot believe that! Here we are at the hotel. I will drop you off at the door, and then go look for a parking spot. See you inside!"

Once inside the door, I immediately spot Jim, waiting in the lobby to greet all of us and direct us to the conference room where we will discuss and come to a consensus as to how this initiative will begin post-firefights. "Good morning, Jim. Thanks for putting together this meeting."

"My pleasure, Liz. I truly want this Lean change management to get off the ground and be successful. The state will not be giving out TARP* funds forever."

"Oh, look! Here are Helen and Nancy, and just behind them are Mark and Tom. I am glad that you invited Tom to participate in this also, Jim."

"Well, let's all head down to the conference room I have reserved for the morning. Liz and Mark, I would ask that you lead this discussion this morning."

"We are glad to do that, sir," Mark replied for both of us. "Liz, how about you explain what we want to accomplish with a new management system?"

---

* T.A.R.P., or Troubled Asset Relief Program, was a taxpayer funded government bailout program of the financial banking system created by the federal government in an attempt to curb the ongoing financial crisis of 2007–2008—to free up credit for small businesses with the ultimate purpose of expanding business and creating jobs.

"Thanks, Mark. First of all, Mark and I have already talked together about the need for focusing on MBV rather than MBO. Does everyone understand the difference between the two? In MBO we are focusing on the performance of the business: did we hit the numbers would be an example. In MBV we are managing by values we have determined will drive the success of the company: things like quality, teamwork, diversity, etc. Does everyone agree this MBV is the way we want to go? Does it make sense to you? If so, let's do it. If not, let's talk about it."

Jim and all of the others nodded their heads up and down.

"OK," I said. "So let's discuss how the old paradigm (MBO) differs from the new management paradigm (MBV) and make certain that we all agree to what we are embarking upon."

Mark began the discussion. "My understanding of MBO is that what gets done and what gets measured is what counts, and we usually measure only financial factors like operating income and sales and profits. However, MBV tells us that 'how' we achieve success is really what counts. It is the 'process' to get there that matters most."

"We do not set out in business to achieve objectives. We set out to satisfy customers. That is what really counts," said Tom.

"I think I see this," said Nancy. "When I 'reclass' an asset, I am not really achieving any long-term successes—for us or for the customer. Taking the ratio return on assets or ROA, for example, we would do better to look at the ratio ROA as a metric (what we measure), but ask how we could better that ratio month after month to improve the business long term. A high ROA really does satisfy customers. For example, if our ROA were, let's say 15 percent, then that is about twice the cost of borrowing money. Think of the new capital projects we could initiate for the expansion of new products the customer wants!"

"That's a great example, Nancy!" I exclaimed. "Financial results are results, but not an end in themselves. You could also say that MBO is short term, and MBV is long term. If we shoot for the long-term goal, the short-term goal should fall into place automatically."

"Liz and I have also talked lately about empowerment," said Mark. "It is difficult to have a person feel empowered in a vertical organization where one has to climb all the way up the ladder to ask, 'Mother, may I?' and then climb all the way back down to execute. A horizontal organization would be more advantageous."

"I agree with that," declared Tom. "I can already see people in the packaging and shipping area embracing empowerment after Mark and Liz were in there fighting fires last week. I am definitely on board with this new management paradigm."

"That's good to hear, Tom," I said. "And so I think you would agree that this new MBV approach would work better in integrated teams rather than in what Mark and I call isolated functions, such as accounting, marketing, engineering, procurement, etc."

"I agree there," Mark replied. "When I walk into Nancy's fiefdom, I do not want to hear some accountant say, 'What are *you* doing here? You don't belong here!'"

"Plus," Jim finally spoke up, "we tend to 'drop' data when we go from one functional department to another. It is sort of like going through a toll booth on the turnpike. If you drop the change, you cannot really go back and get it."

"Good point, Jim," I said. "And I think we also want to recognize the value of doing more with less. This is a reality most people do not understand; however, once they do understand that eliminating waste gets more results more frequently than cutting cost, they really have fun with it!" I just wanted to make sure that Jim would agree with us that this truly was a part of the new management paradigm.

"So is everyone in agreement so far? Again I would ask, 'Does it make sense to you? If so, let's do it. If not, let's talk about it.'"

"It makes sense to me," said Jim. And Nancy, Helen, and Tom all nodded in agreement.

"I agree, also," said Mark. "And so now let us discuss what we want our company core values to be. Again, our values represent what we stand for as an organization, what our people believe in and—most important of all—our values originate from the customer. We need to stand outside our business, in the shoes of our customers, looking in, to understand value. These values will drive our business going forward from today, and be crucial to our success or failure in this new Lean change management initiative. Our manufacturing processes may change over time, but our values will remain constant."

"I believe we have agreed already that quality will be one of our core values," said Tom. "The customer always wants the highest quality product. And I also think that teamwork should be a core value. We do not want, for example, a them versus us mentality between, say, accounting and manufacturing. We cannot create and sustain speed to market if we have continuous dissention among ourselves."

"But why would the customer care, or even know, that we work in teams? I don't get how teams would be a core value driven by the customer," said Nancy.

"Let me give an example," I replied. "Try to think of a time when you purchased some product from a large department store or discount house. Let's say you purchased a vacuum cleaner. It came in a large box and, when you got the box home and dispersed all of the many parts from the box, you found that there was one bolt missing, which meant you were not able to assemble and use this new vacuum. You search through the papers that came with the product, and you call an 800 number for assistance. When someone answers that 800 support line, you spend several minutes stating your problem. The person on the other end of the phone replies, 'I'm sorry, but you have called the wrong department. Let me transfer your call.' And so you repeat your problem to the second support person, who also responds, 'I'm sorry, but you have called the wrong department. Let me transfer your call.' Do you get the picture? This business is organized in departments or what we in Lean call functional silos. But the customer does not care about your departments, or your org chart. To better create value for the customer, the business should be organized in what we call product families—groups of products that run through the same processes or operations, or involve a similar group of customers. These are cross-functional teams dedicated to specific products or specific customers."

"I get it now," said Nancy.

"And I believe that communication and diversity should be core values," said Helen. "After the HR systems of merit model Liz showed to me, communication and diversity must be in our core values. And by diversity I do not mean just one's gender and race. I mean total diversity, including mutual respect for all of our differences—as well as all of our customers' differences. We need to better understand the needs of a wide variety of customers to increase the sales of our products and create new products."

"We have expressed the importance of leadership over management," said Jim. "So perhaps leadership should be a core value."

"More importantly, leadership as a core value implies that the company should be the best at doing what customers say they need, not necessarily being the technological leader. Decades ago an aerospace engine manufacturer may have been the leader in technology as they built the fastest, most sophisticated military jet engines in the world. Then, after the Cold War ended, they were the leader in engine manufacturing—plain vanilla commercial engines, and lots of them. Then, after 9/11, they became the leader in what they called 'Power-by-the-Hour' as they leased engines to commercial airlines and charged them for the engines only when the planes were in the air, earning revenue. Their leadership position in the marketplace changed as the customer requirements changed," Mark explained.

"That's a great example of leadership, Mark, as well as giving the customer what he wants, when he wants it, and at the price he wants," Tom said. "And so perhaps we should add satisfying customers as a core value. Sometimes it is hard for me to satisfy the real customer—the person buying our products—when so much stress is placed on me to satisfy the Wall Street customer, too. I often feel like I am sitting atop a fence, and whichever way the wind blows determines which customer I must satisfy that day—the customer buying product, or the customer buying our stock!"

"If everyone is in agreement with the following core values: quality, teamwork, communication, diversity, leadership, and satisfying customers, raise your hand and say 'aye,' or forever hold your peace!"

"Let the record show that everyone has agreed, and so the next challenge is to decide on the behavioral objectives that would help us implement our basic core values, because the best way to communicate a company's values is through behavior. Who wants to start?" I asked.

Helen spoke up first. "I believe we should have individuals paid based on their contributions to the business. And monetary bonuses traditionally given out at year-end would be replaced with nonmonetary rewards given out in celebration of our team achievements at any time throughout the year. This would also help with diversity if new hires knew that pay is based on contributions to the firm, no matter where you went to school or the number of years of service you have."

"This is a great idea, Helen. It makes me think of Mark's dog training school again. Correct me if I am wrong, Mark-san, but you do preach AAA: 'applaud, accentuate, and award' as a motivator with your dogs. I believe this thinking should spill over into our business world as well. Perhaps in business we should preach CCC: 'congratulate, compensate, and celebrate' as a motivator."

"Yes, Liz, and dismantling functional silos will help with teamwork," said Mark. "We have a lot of goal incongruence with our current department/silo setup. A product family-type setup, where products that run through the same processes or machines are grouped together and people from each functional silo are assigned directly to any given family, would help with goal congruence and teamwork and also the core value of communication throughout the enterprise. Dismantling functional silos would also help the accountant to get out of his 'command and control' mind-set, and become more of a team member with other 'functions' inside his 'family.'"

"Teamwork would also be reinforcing the behavior we want in that an individual is no longer responsible but rather the team is responsible and accountable. People would be more apt to speak up in a team," said Tom. "The fear of making a mistake is now the burden of the entire family—not any one individual. And teamwork would help to ensure the core value of quality as quality 'problems' would now be viewed as quality 'commitments' by the product family. And all of the above helps us implement the core value of satisfying customers."

"Then we just need a behavioral objective that would help us implement leadership as a core value," I said. "Actually, I believe that replacing our current accounting system would help with all of the core values, but especially leadership. For example, Lean tells us not to make one until the customer takes one. Yet our traditional standard absorption cost system tells us that we need to keep all of the people and machines busy 24/7, making product to absorb overhead. So not using this standard cost system in managing the business would help us implement our core values of quality, teamwork, communication, and satisfying customers. That is four out of six core values with just one behavioral change! And, developing and implementing a Lean cost management system would force financial folks into a leadership role. They would no longer be sitting in cubicles, writing history, but rather they would be out in operations, fully engaged in their product family and assisting their family in whatever business decisions need to be made. Then, too, these 'families' promote the core value of diversity due to the mix of functions of people in any given product family."

"It would also help with quality as we would be working on eliminating inventory instead of valuing inventory!" said Tom. And Nancy agreed with that, stating that she hates the daunting year-end inventory task and monthly cycle counts.

"It will also help you, Jim, and the product line managers with leadership as you will have financial as well as nonfinancial info to manage the business on a weekly or even a daily basis as opposed to a traditional income statement which is very lengthy, only comes out every four to six weeks, and takes time away from actually satisfying the customer." This is a point I have wanted to make for some time, and now was as good a time as any. "And, Jim, this product family design certainly improves the diversity of the business as we now have cross-functional groups of people sharing a variety of ideas and perspectives on products, customers, and problem solving."

"You are right," added Mark. "Just last week I went to cost accounting, in need of some data for a continuous improvement project. Nancy, please do not be angry with Carol, but she told me to come back in two weeks. You all were busy with 'the close.' Then, when I did return, all I got was what I would call a 'data dump' of debits and credits. How would you like me to give that to you, Jim?"

"I would not like it at all, Mark. And that is why I have you," Jim said with a smile. "But you and Liz are correct. I do need daily information, not just data, and I need it in a format that is well summarized and easy to digest. I believe you, Mark, expressed to me the importance of replacing PowerPoint presentations with something you called a 'dashboard.' These traditional financial packages that come out about every six or seven weeks are a lot to digest. It may take someone many days to get through it all! It would be great if I could have information on my laptop at any hour, any day, to manage the business."

"You will have just that when we get back and get started on this!" I replied.

"Good," said Jim. "But now that we have our core values, and our behavioral objectives to integrate our core values into the business, how do we introduce these values to the associates of the company? Not everyone will agree with these values, nor will every current associate possess these values."

"I have seen this done with a matrix," said Mark. "We would place every 'like' functional person on one side of a matrix, and then the qualities or values we want in that function across the top of the matrix, and decide who we want to keep for what positions."

"I have an idea which I experienced at a smaller company. The CEO just brought everyone in the business together and announced, 'Starting today, we are changing the way we do business. If anyone does not like change, it would be best to leave now as you will not be happy here.' Thus he downsized the company voluntarily at the very beginning of the change, and saved an HR person like Helen a lot of heartache," I said.

"Did it work?" Jim asked. "I mean it does seem a little harsh."

"It sure did," I said. "There were quite a few people who really did not want to change and accept the core values, for whatever reason, and the group we were left with was really excited to get on with the Lean change business transformation."

"Great!" Jim replied. "Then this is the way we will go. Helen, would you type up all that we have decided on today—from core values to behavioral objectives to meet those core values—and get copies to everyone here. I want all of us to sign off on this model for implementing our Lean change management or 'new management system.' And we will begin tomorrow morning at nine with a town hall! I hope you all are as pumped about this as I am!"

"Yes, sir," we all replied in unison. And I could not help but notice that Jim's chauvinism continued to show, even as he asked Helen to "type up" the minutes of our meeting. Mark has a lot of choices for conversational topics on the golf course.

4

The CEO's Office after Hours

The best way to communicate a company's values is through behavior—in particular, management's behavior. I cannot say that enough times. Then, once a company has articulated its values and thought about how the behavior of its employees will encourage or discourage these values, it must develop specific systems to make those values come alive and change behavior. In other words, "We must live our values!" Many will say that this is not the job description of a controller. However, one of these systems will be our new Lean cost management system.

We had covered core values thoroughly at our offsite meeting, and we now had a small administrative team, which we called the Steering Team, dedicated to making this Lean endeavor successful. Before the town hall meeting tomorrow morning in which Jim would inform all associates that, starting now, we are changing the way we do business, I wanted to talk with Jim one-on-one. I wanted him to tell me, face to face, that he was committed to this Lean endeavor for the long haul.

I knocked on Jim's door, and he opened it cautiously. "Hi, Jim. Are you in a hurry to head out, or do you have a few minutes?"

"I have plenty of time, Liz! Come on in. Please, have a seat."

"Thank you, Jim. I just wanted to touch base to ensure that you are comfortable with our plan to begin this Lean journey—for the second time."

"I am committed to this change, Liz. I will not allow anyone to interfere with this Lean implementation this time. As I told Mark earlier, this state TARP money, along with a large line of credit from the bank, is keeping us in business right now. Both these TARP funds and the bank loans must be repaid. We must get a strategy in place to improve both revenues and profits, have successes, and turn this business around."

"I am glad you see Lean as a business strategy, Jim. Many CEOs only see Lean as a series of tools or stunts, and give up when the going gets tough. You did give up the first time. May I ask why?"

"Actually, when the economy went sour and this latest recession started, I panicked and went back to cutting costs to save cash. I wanted to be certain that we were looking good in the eyes of our bank."

"So, to be short and blunt, accounting was the reason for stopping the Lean initiative the first time?" I asked.

"Yes, I guess you could say that," Jim replied. "In your experience, Liz, why do companies quit their Lean programs?"

"There have been many studies done, Jim, on just this topic. I would say that the number one reason businesses fail or stop Lean is lack of leadership. The second most popular reason for quitting Lean we found to be a lack of vision or values. You might also say that these companies had no road map for where they wanted to go and no plan on how to get there. And a third most popular reason for exiting a Lean initiative would be a lack of passion among the business associates overall. The entire organization has to be fired up and want this, and this happens best when there is a crisis. Yet another reason for stopping a Lean transformation is the traditional absorption cost accounting system. Executives expect to see big improvements in the bottom line the first year, but this rarely happens. Then, too, my own thoughts and experience tell me that organizations fail at Lean transformations because the top managers, especially the CEO and CFO, do not understand the difference between cutting costs and cutting waste. May I ask if you now understand the difference between *cutting costs to save cash* and *cutting waste to create cash*, Jim?"

"I believe I do, Liz. When we cut costs, we are cutting resources, such as people, plants, and equipment. That only saves cash in the short term, and we cannot do this for long until we have no people, no plants, no equipment, and thus no business! However, when we cut waste, such as inventory, we are creating cash by converting that stored material, labor, and overhead into hard cold dollar bills. But I admit I have a hard time practicing cutting waste. I know inventory is a big waste that ties up cash, but when I see a good deal on raw material I cannot help but buy it all up! I know we have buildings full of material and have even purchased trailers to store all of this stuff, but ..."

"I understand your feelings, Jim. Having raised four children, I know I want to buy all the chicken I can fit into one cart and into my freezer when

I see it on sale at the supermarket. It's hard to break old habits—even when our immediate situation has changed."

"I also understand that reducing the waste of inventory is just the beginning of a Lean journey, and that just-in-time is really a tool used in implementing Lean and not one of the principles of the actual business philosophy," Jim admitted.

"Yes, we will want to implement each principle of Lean, not just the tools, such as JIT. If we can focus intently on, and successfully implement, just the first three principles of Lean: creating value for the customer, mapping the value stream, and then creating one-piece flow, I believe we will have a good foundation in Lean that can be sustained. And best of all, we will have changed people's behavior here. Do you have any reservations, Jim—anything in particular that you see that you fear may impede this company's second attempt at Lean?"

"This trip my main concerns are with certain personalities and their behavior. How do we change behavior, Liz?"

"You are correct to be concerned in this area, Jim. For starters, I would comment that most engineers, who typically lead Lean transformations, believe that change is a technical problem, but really it is a behavioral problem. Mark understands this! I must admit that I have learned a lot about behavior from Mark and his dogs while I have been out on his ranch. Identifying or discovering what have been concealed capabilities and talents in dogs or people, arousing what were once stifled passions, or finding a new outlet for what is obvious 'get-up-and-go' and high energy, methodically changes any previously learned behavior. As Mark would say, 'the correct incentive, the correct game, the correct benefit for service!' is what changes behavior. But, first of all, we need to communicate to everyone in the town hall tomorrow morning that we have a serious crisis here. I know you do not like to reveal financial information to our associates, Jim, but you need to let them know that right now the state is paying their pay check—not the customer. That is always the first step: create a crisis. We do not have to create a crisis. We are living one! Let them know that the business may not be here next month if we don't change things around here."

"You are right, Liz. I do not like showing financial data. You are big on visuals. Perhaps you could help me with some graphs that would vividly show the big picture without giving people a lot of numbers."

"Sure, Jim, that's a great idea!"

"But then, too, Liz, how will people know I mean business this time? They have been through so many 'programs-of-the-month'—from reengineering to TQM to JIT, etc. How do I communicate that this time is different from the last? You gave me several good reasons why businesses fail at Lean, but I see our potential reasons for failure this trip as: too much resistance to change, both in OPS and in management; too many previous reorganizations; and we are out of cash! Is it not too late to communicate any more improvement programs and remain credible?"

"You need to focus not so much on communication but rather on leadership this time around, Jim. We have talked about this before. We do not need more managers; we need leaders! Be passionate, Jim, and emotional. Show them how what we are doing now will not lead to a successful business. Elaborate on our core values that will guide us. Assure them that our value of diversity means we value viewpoints and ideas from everyone—the customer as well as the employee. Involve them, Jim, and ask them for suggestions. Don't treat them like 'Lucies' in the chocolate factory! Tell them of your vision and strategy for the company. You've shared with me your strategy to take the cash from reducing inventory and use it to grow our product line from our niche market into other new markets. Tell them about this! Paint a picture of your vision for this company ten years from now! Tell them we will empower them by teaching them to work differently and show them how to manage change by tying all the parts of the business together."

"Can you give me an example, Liz?"

"Sure! Look at what happened in packaging and shipping just recently. Everyone has heard about it. Some buyer in purchasing wanted to make sure he had no negative purchase price variance and got a bonus, so he bought the cheapest boxes he could. When the bottoms fell out of the boxes, it created havoc in operations and cost them money. But did the buyer care? Another example would be our habit of keeping design engineers in one location, and manufacturing engineers in another location. The design engineers design a product that is high tech, but something the manufacturing engineers cannot make. How does this happen? They do not sit together nor talk to one another."

"This all sounds good, Liz, but perhaps I should just send out a memo to everyone."

"Absolutely not, Jim! A memo tells people that we, management, are busy doing it all, and will let them know when dinner is done! We want to involve people. You can't empower people, then not give them the knowledge they need to be successful. Tell them about the five-day Learning Kaizen in which everyone will participate. Tell them that the first two days we will be giving them instruction on the principles of Lean, not just the tools, and the last three days we will be asking them to make suggestions for improvement, and then implement those ideas. No more boring book clubs! These five-day improvement sessions will be Lights! Camera! Action! Everyone will have the opportunity to pitch his or her solution to problems to management on the fifth and final day of the Learning Kaizen. This then reinforces our core value of communication. And, at the end of the town hall meeting, you will need to take some questions from people. Are you willing to do that, Jim?"

"Yes, and I see where you are going, Liz. Once people see that we are listening to and implementing their ideas, behavior will change. But will the behavior of some of my management team change? Two of these people really need to accept each other's views. I have even thought of sending these two people to a psychologist."

"You do not need to send any of your management team to therapy sessions, Jim. Once we have created *value* with product families and created *flow*, and have these two Lean principles in place, and we have a new Lean cost management system inside the product families, you will see behavior change. The old behavior will cease to exist."

"I hope that is true, Liz."

"But you also need to 'walk the talk' and lead by example, Jim, so I have asked Mark to place you on one of the first Learning Kaizen teams. For the first time ever in the history of this company, people will not be talking about 'checking their brains at the door.' And we will all soon be working in teams full time—not just for Learning Kaizens. No more Lone Rangers! We have core values that we and our customers share. You have a vision that you have shared, and a plan to get there. No more goal incongruence. No more hammers looking for nails! We are all essentially on the same team. Our Learning Kaizens will allow people to voice their ideas, both positive and negative, without fear. Your town halls will be a sharing of information and updating of the vision and plan on a regular quarterly basis. No more lining people up after the annual town hall and mowing them down with machine guns. We have communicated that we are all here to stay. And, if they should think we are just talking heads, wait until they see the furniture start to move and smaller permanent teams develop!

These smaller teams, our *product families*, and what happens inside our small families, will ensure the correct behavior in all of our associates, as well as ensure that we sustain our Lean improvements this time around, Jim. I even have a visual for that!"

"Like I said, Liz, I have complete trust in you and Mark. You've helped me just now. Help me with some graphs, please, and do keep talking to me. But please answer one more question for me, Liz. Since we started Lean the first time, why did our culture not change?"

"Because culture is day-to-day behavior, Jim, and we did nothing to ensure that our associates' day-to-day behavior changed. We have been organized vertically, by function, in a 'mother-may-I' type of organization, but value flows horizontally. We need to realign people to change behavior and thus change the culture. This works best in cross-functional teams in which people have been co-located. Right now, in our silo organization, accounting is not talking with OPS, which is not talking with procurement, etc., etc. This time we are setting up the redesign of the organization so that we can create a system. We are creating a new Lean system with all of the pieces linked together, and we will succeed—with the help of the new Lean cost management system."

5

Henry Ford, the Father of Lean

"Industry exists to make things that people use. But when managed by men who know nothing of the factory, whose interest is confined to the balance sheet, its principal products become dividends.... When the chief function of any industry is to produce dividends rather than goods for use, the emphasis is fundamentally wrong.... Business never fails; only men do that."[1]

The town hall meeting the next morning to announce the change in the way in which we will do business went very well. The response by our associates to what Jim had to say was most positive. Jim had everyone's attention as this was not the usual end-of-year town hall meeting where masses of people were let go. There was no lack of leadership or passion on Jim's part. One could hear the emotion in his voice. Our people did ask questions, which I thought was a very good sign that the fear was beginning to dissipate, and Jim responded well to those questions.

"What about layoffs? Will there be layoffs?" Of course, we all knew this would be the very first question asked of Jim.

"In the past, we have had town halls only at year-end, and some of our people were always laid off at this time every year. We now understand that laying off employees is not the way to get productivity gains. No more! From now on, the town halls will be held quarterly to celebrate our successes and talk about what comes next. With that said, some of you will decide that change is not something you can do. For those of you who want off the highway of change, here is your exit ramp! For those of you who will commit to this new philosophy of doing business, I promise you that the team today will be the team next year and the year after that. Yes, we may need fewer people as we work faster and smarter, but then we will utilize our best people as leaders and teachers and facilitators—not lay them off. As we grow the business, we will need people who know the business inside out to act as coaches, Learning Kaizen leaders, product family managers, and mentors to other members of the company. You will understand this more tomorrow when we start the Learning Kaizens. These five-day improvement and learning teams will give you the knowledge of the Lean principles that you will need to succeed. We want everyone to succeed. We need this business to succeed."

"How did we get into this mess in the first place?" was another question asked by an associate up front in the room.

"We are in this crisis not because we did something wrong. No one is at fault here. We are in a crisis because we outgrew the tools and philosophy of our business, and our values were not aligned with our customers. Until yesterday, we didn't even know what our values were, or who our customers are! Today we have core values, and we have a business strategy, which is to reduce inventory and use the money we create to acquire more business, thus increasing both revenues and profits and growing our business into new markets. We are in a new age today, and yesterday's manner of solving problems will not work today. We are also borrowing money from the state and the bank, and we cannot sustain business this way much longer. We must change the way we work, and change the way we think. We need new ideas, and we need to become a learning organization. We will be asking you for your ideas."

"What exactly do you mean by 'values' and 'change'?" asked another employee.

"Value is what the customer is willing to pay for. Quality is an example. As for change, we will be changing our focus—focusing on our customers and on you, the employee. We need to listen to our customers, and we need to listen to all of you. Then, too, we will change how the business is structured. We are going to tear down our departments and pods of office cubicles, which prevent people from working together and talking with one another, and build *product families* in which representatives of each department will come together as a small, manageable team. We will start with just one product family as this is an experiment or pilot program. We will give you the knowledge, and then ask you to put this knowledge into action inside your family on the shop floor. Liz and Mark will be telling you more tomorrow when the Learning Kaizens begin. Liz will be explaining this philosophy of business to all of you, and Mark will help design the setup of the first product family. After the first product family is in place and working well, we will introduce other product families."

"What do we have to improve? How do we know what to improve?" someone asked.

"You will hear the word *system* a lot from now on," said Jim. "Our business in the past was just a lot of departments where the people in one department didn't know what the people in the other department were doing. Starting now we are going to link everything in our company into one system. We will start by linking knowledge with action in our Learning Kaizens. We will eventually link our customers and suppliers outside of the company. But inside, for now, we will link OPS and management, and OPS and accounting, and OPS and procurement, and we will link what we call QCD—quality, cost, and delivery. We do well on delivery, but we also need to incorporate excellent quality and give the customer what he wants at the price he wants. We need to execute all of these simultaneously. Presently the only system we have is a system of stunts! You might say that our future system will be a system where we all share in the work and the rewards."

"Well, we tried this once before. How is this time any different?" shouted someone in the back of the room.

"The first time we tried Lean, we focused just on the tools and not the principles of Lean. Reducing inventory and conducting kaizens are examples of Lean tools. Creating value for the customer, mapping the value stream, and implementing flow are examples of Lean principles, which will be our focus this time around. Liz will be explaining all of this to you in the context of Henry Ford and his business years ago, for it was Henry Ford who instigated this commonsense approach to business. Then, too, when we attempted Lean before, we conducted training sessions, but we never linked that training to action, and we never carried through to create the Lean principle of flow. We also caved in when the recession kicked into high gear, and we went back to cutting costs, which is not the Lean philosophy of business. Liz will be explaining more of the Lean principles tomorrow morning. Ask her any and all questions you may have! Let's all look forward to learning how to manage our new business system to deliver exceptional service to our customers and reinvent this company.

"Liz will take the entire group for the initial day of training on the Lean principles and creating value. Then Mark will facilitate the second day of Kaizen Learning to teach the concept of VSM (value stream mapping) and how we create flow. The third and fourth days we will break into our small teams, which Mark has already set up, and have our teams put into action what they have learned by implementing 5S in their own areas. Mark even has the members of the management team in these small learning improvement teams, and one team will be doing 5S in my office! Today, in this town hall, we wanted to communicate the strategy of the new business, which is to lower inventory and lead time, and then use the cash created from inventory reduction to do mergers and acquisitions to place additional business in freed-up capacity. We will grow revenues as well as profits, and expand into new markets. Tomorrow we shall begin to move the furniture for our first product family! This is where our Lean journey will really begin."

The next morning I got started with the entire workforce. I recalled the facilitator who ran the one-day session on Myers-Briggs years ago, and tried to emulate his energy and excitement. Our new Learning Kaizens would not be at all like the "hammers looking for nails" or the "Lone Rangers" which we had in our first attempt at Lean. One cannot improve a business if one does not first have the knowledge on how to do so. The first two days of the Learning Kaizen would be the imparting of the knowledge, and for my part I planned to use Henry Ford as an example throughout the day. I wanted to emphasize that Lean is not a "Japanese" concept, and also that Lean is not a "fad." Rather, Henry Ford is the "father of Lean," and Lean has been around for a hundred years!

"Good morning everyone! As we begin the first day of our Learning Kaizen, I want you to know that you may ask questions at any time during the day. This is not a class lecture. We are here to impart knowledge to you, which you will then utilize and put into action, and the best way for you to gain the knowledge is to ask questions. If, for example, you do not understand any particular piece of information presented to you today, please do raise your hand and ask for an example or an everyday illustration. Yesterday our CEO told you of his vision for this company, which we are reinventing starting right now. That vision is to grow the company with new lines of business, which we will acquire with the cash from the reduction of inventory, and we then drop that new business into the freed-up capacity we create by working smarter. Henry Ford also had a vision back in the early 1900s. He had a vision for something he called a car, which he would build in a manner such that every man could afford to buy one. Thus he was providing a 'service' for people. Do any of you know anything about Henry Ford?"

"I know he made the Model T. And he was the one who started the Ford Motor Company," someone responded.

"That is true, but do you know anything about Ford the person? For example, did you know that he was born and grew up on a farm in southern Michigan? He was a farmer—just like many of you! He was one of many children, and did not have an easy childhood. His mother died when he was just twelve years old and, since he had been schooled at home or in a small one-room school house, his education did not formally continue after the age of 17. But he did go to work as an apprentice, and built his first steam engine when he was just 15 years old. Then he worked for Tom Edison. Can you imagine how exciting that must have been to learn hands-on from Tom Edison? As a young man, others who were close friends of Ford were the Wright Brothers (Orville and Wilbur), William Harley, and Arthur Davidson. How cool is that? The story told is that these guys would get together on a Saturday night for beers, and talk shop. I know I would have loved to have been a fly on the wall on just one of those evenings!"

"That would have been fun! Tell us more about Henry Ford, Liz."

"Well, the story of the development of Henry Ford's 'common sense' philosophy of business is an interesting one, and I do want to share that with you. Like I said, he had an idea or vision to provide a service for people with a new product he called a car. He would create value for his customer by making this car affordable for every man. He designed and built this car in the basement of his home. Yes! In his basement! Just in case this car would be successful and people would want more of these cars, he wrote down in a book every step in the making of this car. When the car was successful and everyone wanted one, he had two choices. He could hire one person, teach that person all 500 steps in the manufacture of this car, and they could then make two cars each year! Or, he could hire 500 people, training one person on each step in his book, and he could then make many cars each year. Thus, Ford became the father of what we call the 'atomization of work.' As he could not fit 500 workers in the basement of his home, he bought a large facility and set up the first moving assembly line in the first manufacturing plant. Of course, it only made sense to Ford to set up this assembly line in the order in which the operations occurred in his book. This is what we Lean practitioners today call *flow*—one piece at a time, placing value-added steps end-to-end. This is what Jim, our CEO, was referring to when he told you that we would be starting our Lean initiative by 'moving the furniture.'"

"I still do not understand what you mean by 'value'?" asked one associate from the factory.

"Let me give you an example. Not long ago I met my son in Dallas. I flew from Tampa to Atlanta, and then a second connecting flight took me from Atlanta to Dallas. That was not too bad, right? But then, for the trip home, the airline transported me from Dallas to Detroit, Detroit to Atlanta, and then finally Atlanta to Tampa. Was this airline creating value for me, the customer? Absolutely not! They were making the best use of their assets (their hubs) and not even considering me, the customer. If it does not add value for the customer, it is waste, and we want to eliminate waste. How do we do this? We talk to the customer! We stand outside the business, in the shoes of the customer, looking in, and ask if the customer would pay for what we see. If not, it is waste and we need to get rid of it."

"That is also a very good example of what waste really means in the context of a business, Liz," commented an employee in the group.

"Yes, it is, but we cannot really eliminate waste and create value for our customer until we create what we call *product families*. Let me give you another example. As I travel a great deal of the time, I have a security system with surveillance cameras on my home. One night before retiring at the hotel, I pulled up on my laptop the surveillance camera on my front door. I saw two kids approaching my glass front door with what looked like sticks or pipes. Then I see them strike the door! Not wanting to wait to find out if the alarm works, I pick up my cell and dial the monitoring company for my security system. After explaining the situation and that I need them to call the police immediately, the person on the other end of the phone tells me, "I am sorry, but you have called the department responsible for updating your personal information. Let me put you on hold while I transfer your call to the dispatch department."

"No! Wait! Do not put me on hold! I have an emergency! Wait …"

"Do you get the picture? This business is organized in *departments*, or what we in Lean call *functional silos*. But the customer does not care about your departments or your org chart. To better create value for the customer, the business should be organized in what we call *product families*—groups of products that run through the same processes or operations, or involve a similar group of customers. If your business is still organized in departments, you are not ready for Lean. Ford thought it was only common sense to line up all of his machines in the order in which the operations occurred in his book of steps, and then asked, 'Does it add value for the customer, or is it waste?' Ford did not push his raw material from material stores to the point of use, nor did he push work in progress from one functional department in one end of the building to another functional department at the opposite end of the building, and so on and so forth."

"That does make sense, Liz. We all do that right now, and then many of us are waiting hours for materials or work in progress to arrive at our work area."

"We are organized in departments which, even if not physical, do not allow information to flow freely. If you see someone from finance or HR, for example, in your operations department, you immediately think, 'This person does not belong here!' Then, too, just think of all of the waste in time and transportation when we move raw material or work in progress from the northwest corner of the building where we have our materials stored to the east side of the building where machining centers are located, then to the southwest side of the building where all welding is done, and then to the northeast section of the building where assembly occurs. If we were to map this movement of material while we are structured in this way, our map would look like a plate of spaghetti! Henry Ford saw this, and decided it only made sense to line up all machines in the order in which the steps or operations occurred rather than structure his company in functional departments. Think of the time and money he saved just by eliminating all of that waste in movement of material and work in progress! Even today, modern, world-class auto manufacturing plants receive raw material into many small receiving docks close to the point of use. Just think of the waste in material handling alone, and the money that is created by the removal of this waste. All they have to do now is slide that material over to the assembly line!"

"That does make sense, Liz, but what about the material handler? Is he now out of a job?"

"Absolutely not! We will better utilize him somewhere else in the business. First we are going to tear down all of our functional departments, and create *families*. Each family will consist not only of associates from each silo but also it will be made up of similar products that run through like machines or like processes. In each family, we will have at least one person from every department. That is to say, we will have at least one engineer, one financial person, one quality person, one procurement person, and so on, in each family. And each family is totally accountable for that family's operations and costs—just like your family at home!"

"I like this family idea of structuring a business, Liz. It obviously made sense for Ford, and it makes sense for us here also."

"We may keep some product development engineers outside of the families, as well as a small group of facilitators for Learning Kaizen events. Then, too, our company leaders such as the CEO, CFO, and COO will sit outside of the families, as well as the shippers who move from family to family picking parts for shipping. Our commodity managers will stay in the field, working with our suppliers on behalf of the entire organization, but buyers will sit inside of a family. And the sales team will be in the field also, thus outside of the family from a cost perspective—although each salesperson may be assigned to the family whose products he sells. Now think about this, for just a moment, from a financial viewpoint. No longer will you in OPS have all sorts of costs allocated to you. All of the costs in your family will be direct costs—material, labor, supplies, supervisors' salaries, utilities, outside vendors—all will be costs that you and your family are directly responsible for and in charge of. They reside in your family, and you are accountable for them. For the first time ever, your P&L will be yours to control!"

"That does sound great for us as well as the customer, Liz! Last month someone in accounting yelled at my OPS department because we were not helping to make the company better, but then more than half of our costs were allocated to us. We have no control over what the company pays the executives or the legal beagles or the research and development guys!"

"You are absolutely correct and justified in feeling the way you do, and this is all about to change. Of course, creating value and mapping that value stream are just the first two principles of Lean. The third principle is creating flow. Henry Ford did this when he lined up all of his machines in the order in which the operations or steps in making his car occurred, eliminating the waste of material handling as well as any other waste he saw in the line. He was smart and a visionary, and wanted no part of any plate of spaghetti in his plant!"

"He does sound like a cool guy. And you say he was a farmer who had little schooling after age 17? Whoa! This guy is impressive. He must have learned a lot from his famous innovator friends, or he was just smart to begin with!"

"I believe he was definitely about 50 years ahead of his time!" I commented.

"I'm pretty sure I understand the principle of creating value for our customers after your examples, Liz, but can you give us a modern-day example to illustrate flow?"

"Sure! I want you to think of a big wedding or graduation or birthday party for which you had to send out many invitations. You probably had the invitations printed and purchased envelopes for those invitations as well as stamps, and you probably made up address labels so that you did not have to do all of that handwriting on the envelopes. Now you have a choice in how you can put together these invitations. You can pick up each invitation, place it in an envelope, put on the address label, and then the stamp: that constitutes four *processes*. Which do you believe would be more resourceful: doing all four processes for each of the 50 invitations one at a time, or would it be faster to do one operation for all 50 invitations, then the second operation for all 50, and so forth?"

"I think I would rather have someone help me, and do what Ford did. I would have one person stuffing, one person putting on the address label, one person stamping, and move those invitations one at a time down the assembly line."

"Good choice!" I remarked.

"Liz, I have another example, I think, where my kids are creating flow when we plant our big garden each spring. Our older son is in our large cold frame, and hands out each seedling in its carton to my younger son who places the seedlings in one of the children's wagons. My older daughter is the material handler and pulls the wagon out to the garden, taking each seedling out of the wagon as my younger daughter plants each small carton in the big garden. Is this what you mean by flow, Liz?"

"Yes, it is! And perhaps next year you could build a cold frame closer to that garden, then utilize your daughter who is the material handler for other necessary work, such as helping her mother with the canning of those garden veggies." Everyone laughed.

"And then there is a fourth principle of Lean, which we call *pull*. Stated simply, Lean says we will not 'make one' until the customer 'takes one.' In other words, we will not build inventory. That would be a waste. Just think of all of the money in materials and labor that are tied up in inventory that we cannot sell. The best example of pull that I can think of might be the Pizza Hut Express in the airports. Have you ever been in a large airport where there was a Pizza Hut Express?"

"Sure!" someone hollered.

"And did you see them put out 500 cheese pizzas, 500 pepperoni, and another 500 supreme pizzas at any one time?"

"No, I sure did not."

"That's right. They put out just a couple of each variety of pizza and then, when the customer takes one, the pizza express makes another one of that specific variety."

"Do we have to implement *all* of this in our business? This sounds like a lot!" someone asked.

"While the implementation of Lean may take several years, the successful implementation of just the first three principles (creating value, mapping the value stream, and creating flow) can change our business and make it better in a short time—and keep it successful while we work on the other principles later in our Lean journey. To minimize the chances of becoming overwhelmed, we will implement this one family at a time! Does anyone have any questions on creating value for the customer, and establishing flow? If not, to get us started, we are going to break into small groups, as if we were small families, and practice utilizing value and flow to establish 5S—a Lean tool I know you are all familiar with from years ago. In the past, 5S was just a 'command' to 'Clean your room!' But this time we are implementing 5S as an example to teach how we implement the Lean principles. Mark and some other kaizen facilitators will lead tomorrow's events. At the end of the event, each team will have the opportunity to present to management what they have accomplished. And even our CEO and CFO will be participating in a team. We are all in this together! Then we will begin our Lean journey—one product family at a time. The members of this first family will be announced at the end of the week. Mark and some of you in the family will 'move some furniture' to create flow. Then this first family will become the model, and the family's members the trainers for the next family, and so on and so forth. Are there any questions?"

"I have a question. Is management really going to listen to our ideas? Are we really free to speak up?"

"Absolutely! You heard our CEO say yesterday at the town hall meeting that everyone would hear the word 'system' frequently from now on. I am surprised that none of you have asked about the meaning of the word 'system' in the context of Lean. Henry Ford probably thought about systems as a small boy when he repaired watches. All of those gears working synchronously denote a system. Then, too, think about a social system. Your church is a social system. Your neighborhoods and local schools are social systems. So why should our business not be a system? We should all be sharing in the work and the rewards, just like at church or school. But do we do this in business? Do we listen to people in all parts of our business? Probably not in the past, but we will in the future—starting right now."

"Can you give us another example, Liz?"

"Sure! Did you read in the paper recently about the town that was experiencing severe spring floods? The water kept rising and rising, and a few people finally decided that the only way to save their homes in one part of town was to blow up a levee, and allow the flood waters to flood another part of town where the farmers had planted crops but where there were no homes. We also refer to this as *goal incongruence*, where one part of an organization does not care what happens to another part of the organization. We do not want that in our Lean management system! I like to call this old type of continuous improvement 'hammers looking for nails.' In other words, we go around looking for what we can do to improve our own neighborhood—not realizing that we are part of a larger ecosystem and are really hurting the whole system and thus ourselves eventually."

"Thanks, Liz. Those are good examples."

"So let's wrap up what we have learned today. The first three Lean principles are: create value, map the value stream, and create flow. Basically what Henry Ford was telling us is that what are most important are successful human relationships—what I call behavior. Thus creating value in multifunctional product families requires some new learned behavior. Then eliminating waste in the value stream and creating flow by lining up operations in the order in which the activities occur also require successful human relationships and cooperation among people in teams or families—more new behavior. Henry Ford believed that failure is simply the opportunity to begin again, this time more intelligently. And we will do just that!

"So we are beginning this Lean journey again, and this time more intelligently as we are employing Lean principles—not just Lean tools—and we are looking at our business as a system. Ford believed that the worker himself was the company's best asset—as well as its best customer! Your executive team here today realizes this. Ford believed in adding value. He stated that middlemen (i.e., dealerships) did not add value to his auto business, and told his workers that he was producing cars to sell—not cars to store. Ford saw workers, not as commodities to be bought and sold, but as his best customers. Your CEO also sees this. Ford believed that his Lean principles could be applied to any business, and he did apply his principles to successfully run a hospital, a railroad, and a school. So we, too, can do this in our business! Ford told us in his writings that there is no recipe or cookbook for a successful Lean implementation. We will write our own book as we go along—learning by doing."

"This is a powerful, as well as a simple, business model, Liz."

"Yes, it is. Tomorrow Mark will lead you through mapping the value stream and more on creating flow. But the real start will begin after the creation of the first product family, and the moving of the furniture. We will learn by doing, and never stop learning. Let's just do it!"

REFERENCE

1. Ford, Henry, and Samuel Crowther. 2003. *Today and Tomorrow.* New York: Productivity Press, pages 215, 252.

6

Let's Get Started! Surfing atop the Breakers of Change

> The universe is a system. Government—be it local, state, or federal—is a system. The plants and creatures in my local community, as well as yours, are an ecosystem. Any large body of people (or creatures and plants) is a system. So why is it so difficult for a business to see itself as a system? If only a business today could understand and see itself as a system, it could survive and thrive in any economy.

I hope that I have made my points so far to Jim and the others on the Steering Team. The key words in our endeavor to successfully implement Lean change management this second time around are *behavior, value, flow, system,* and, soon to be introduced as the sustainer that will glue all of this together, a *Lean cost management system* (LCMS.)

The initial Learning Kaizens went well, and the Steering Team is meeting once again to establish the first product family. Mark has chosen, as members of this family, persons from each department, including a cost accountant, Carol, from the accounting department. He has chosen Mindy as the product family leader. And all members of this first family have been chosen because they are persons who can influence others.

We argued for some time over the value stream for this product family. Can the value stream pass from one floor to another floor in a building? Can the value stream pass from one building to another building? It was difficult to convince Jim that the value stream for this family, our largest product line, did not have to stay within the boundaries of any one building—but that this situation was not optimal. Typically, the value stream does not have to stay within the boundaries of the plant, but that is what we call a "macro map" and we'll save that for a future date. The particular product line or family we have chosen to work with first travels from one floor in one building to a higher floor in that building, and then to three floors of a second building. And in between these two buildings, the product or work in progress (WIP) takes a long sightseeing journey on conveyors and even on trucks, traveling for miles and miles and days and days, apparently doing nothing but elongating the lead time.

Our first product family decided to name itself the A-team, after Henry Ford's very first car, the Model A. This product family is now a system rather than a collection of disconnected people, actions, transactions, and tricks on a highway filled with toll booths. And the family's job, should they decide to accept the challenge, is to make this family a Lean system, where the focus is on value, flow, and thus, cost management, effectiveness, and problem solving.

Once we all agreed on the current value stream map for this product family, Mark went to work "moving the furniture" or what he called "right-sizing" the value stream for this product family. By right-sizing Mark told us that he has miniaturized or designed the value stream for the production of just this one product line for the best possible "flow." We still have one monument, as our corporate Lean sensei Mr. Sato calls it, in our family, which serves all the families and will have to be replaced with a like but smaller machine—one the size of a toaster, as Sato-san would tell us—in each family that needs one. All other furniture has been moved so that the material flows in the order in which the operations occur. We want no plates of spaghetti here!

I have not said much to anyone yet regarding our new cost management system, but Mark's operational redesign for this product family will be the foundation for this new cost system. I will allow people in the product family to discover this without being told as our Lean journey this trip is about learning what works for us, and not following any one recipe. Also, this new LCMS will come into being after all of the physical changes have been implemented in the plant, that is, the moving of the furniture to create flow. For now, we have our first product family and thus have defined some value for the customer. Mark has helped us map the value stream, and create flow to the best of our ability. Now we bring in the members of this family, and stand back while progress unfolds!

Mindy will make an excellent product family manager. She knows the business inside out and she is well liked by her peers, and thus can influence them. She is not an engineer nor an accountant so she has never been taught how to "play the game" or how to manipulate numbers in spreadsheets or on balance sheets.

The cost accountant chosen for the A-team, Carol, is the associate next in line to succeed Nancy as CFO and a good friend to Nancy, so she will undoubtedly influence Nancy. And she has volunteered for this! Getting her to relocate from her ivory tower to the shop floor and co-locate with her family will be a task that I imagine will take some time. The accountants still believe that "doing Lean" is something left for the final few minutes of the day, if time exists after all other responsibilities (mainly paying bills and "closing the books") have been completed.

Carol is busy with measuring the footprint or the occupied space of the A-team in the plant, as well as gathering depreciation info for all the equipment in the family. Other members of the family are engaged in a treasure hunt for the family's inventory, and organizing it all in one place. They found these raw materials in many buildings and trailers throughout our campus. Others, under Mark's guidance, are busy collecting the data for the current value stream map. These data will be different from the usual financial data, and will include things such as number of people, output per person, throughput time, cycle time, equipment uptime, etc. All of the above, along with our new LCMS and new Lean metrics which we have yet to decide upon, will be the baseline for our new family's progress from which trends will be derived and cost savings calculated. Having had only financial data to measure our business in the past is just one more object that has contributed to our associates' dysfunctional behavior. All of that will change with the new measures in our value stream map, the new LCMS, and new Lean metrics. These new Lean measures and our new LCMS in the product family will also be a replacement for what Mark calls our traditional "control system." Again, our new system design must come first—before a new cost system. Then, with the new LCMS in place, no one will be able to derail our newly created behavior and the speeding train of Lean progress.

There was one very important "heads-up" I wanted to be certain Jim understood that we had not touched on the other night in his office. And so I headed back to Jim's office again at the end of another day.

"Hi, Jim. Got a minute?" I asked.

"Sure, Liz. Come on in." I did not have to knock this time. His door was open.

"There is one more thing we should have discussed the other night when I was here. I don't know how I could have forgotten to mention this, or ask you about it."

"What is it, Liz? It must be important by the sound of your voice."

"It is, Jim. As any business begins to implement Lean, most traditional CEOs and CFOs expect to see results in the traditional financials, and by that I mean they expect to see number results such as better operating income and more profits. They expect to see improvements to the bottom line immediately. Did anyone inform you, at the start of your first Lean attempt, that you probably will not see better traditional financial results the first year in Lean, and why?"

"Well, no, not really."

"So may I conclude that this lack of 'number results' was not a factor in your decision to stop the Lean initiatives the first time around, Jim?"

"No, Liz, we really did not get to the end of the first year the first time around. Pleasing our bankers was really the reason for quitting Lean the first time. But, please, do explain what you mean by 'number results.'"

"As you know, Jim, one of the very first big initiatives in most Lean transformations is to greatly reduce the waste of inventory. As we consume our existing inventory of materials and WIP, the amount of direct materials we need to purchase in the current period is far less than what we purchased in the last period. If we assume that our beginning inventory is the same in this comparison of mass production versus Lean production, and we assume that direct labor and indirect manufacturing costs remain the same in both mass and Lean production, then procuring fewer direct materials and reducing our existing inventory will cause our ending inventory to be much lower. Thus total costs will be higher in the Lean production. Assuming that revenues are the same in both mass and Lean productions, we could easily show a pretax loss in our traditional Lean production income statement the first year. Of course, an increase in cash flow is realized by the reduction in inventory, but the top execs are focused primarily on the balance sheet and the income statement."

"I understand, Liz, and thanks for bringing that to my attention."

"You are welcome, Jim. So, you see, our focus the first year—and we will get to the end of the first year this time—will not be so much on numbers or financials. We know we will probably see a loss in our traditional income statement. But the main focus this first year, Jim, will be on linking all the parts of this business and creating a new system. That is truly the first priority, along with getting the financial folks to support OPS and our new product families."

"Again, I understand, Liz. But this may be easier said than done."

"Mark's new redesign and the new system with product families replacing departments will automatically create a new Lean cost management system, where both cost—the cost of people, material, tools and equipment, utilities, depreciation, outside vendors, and so on—and operations are now linked within a product family. Linking cost and physical operations will occur in each product family, and keep this business in total alignment. This new Lean cost management system will not look like any traditional accounting system you have seen before, Jim, but it will give you the information you need to manage the business and make decisions in real time every day. And it will also change behavior because now people in the product families will be held accountable and responsible for 90 to 95 percent of the costs associated with their family. But rather than explain this in detail to people now, let's see how this unfolds in the first product family, the A-team. I do believe that people will be discovering this for themselves."

7

Accounting's Glass Slipper Does Not Fit

Accounting, as well as HR, will be the greatest obstacles—the crucial and foremost limitations—in our new Lean management system. Yet, in leading the Lean change management endeavor, engineers most often overlook these two functional groups while focusing on OPS and the factory floor. Why? The product families (which create value for the customer) have not been formed, and the accountants and HR people have not been moved out of their ivory towers and included in one of the cross-functional product families on the shop floor.

We're transforming from "batch" manufacturing to one-piece "flow," so it seems obvious to me that the financial standard cost system we are using for batch manufacturing will not work for Lean one-piece flow. But the main reason we cannot use our traditional financial standard cost system in our product families is that we need an "internal" system for "managing" the business as opposed to an "external" system for merely "reporting" the value of every piece of inventory and COGS (cost of goods sold). Even financial professionals have difficulty understanding the distinction between financial and managerial accounting. If every accountant were a member of the army, the financial accountant would be the guy or gal who goes out after the war has ended, counts the dead bodies, and reports those numbers to the general. The managerial accountant, on the other hand, would be the one who is in the trenches with the troops, fighting to win the war.

We're using a standard cost financial system right now that does nothing to help OPS people manage the business day to day and win the war. We accountants yell at OPS people for not "making the numbers," that is, the "bottom line," but we do not give them any suggestions on how to improve. Why? Because the standard cost system does not give even the accountants such information.

"I'm very upset!" Nancy cried. "This new product family business organization is going to create a lot of work for my department!" Nancy was standing in Jim's office, speaking to both Jim and myself, and her head was obviously still back in the old business structure of departments.

"What do you mean, Nancy?" Jim asked.

"We have general ledger accounts set up for every cost center in every department, and that is how we report the business every month. Now I will have to change every single G/L account to reflect a product family. This is too much work!" She was screaming.

"Nancy," I said, "you do not have to map a traditional G/L account to a new Lean G/L account. We will still be using the traditional standard cost system to report the business to our external customers such as corporate and our investors."

"Then you're telling me that I have to develop yet another cost system to report this Lean business?" Nancy was still shouting.

"No, Nancy, each product family will develop its own Lean P&L within its own small business to assist itself in managing its family business."

"How will they do that?" Nancy asked.

"It's simple, Nancy. They know the people in their family, and can get wage records from the HR member of their family. The buyer in each family knows what he is procuring for his family and at what price. And, by the way, this Lean P&L is interested only in actual costs and trends—not standard costs and variances to standards. This is one major distinction between your traditional income statement and each family's Lean P&L."

"People who know nothing about accounting cannot just make up their own P&Ls!" Nancy remarked angrily.

"They will not be 'making up' anything, Nancy. They will be collecting data—and not just financial data—that will enable them to manage their own small piece of the business on a day-to-day basis, and not just once a month. You must understand, Nancy, that the Lean P&L is not a formal accounting income statement. It is small and simple, and goes together fast. And it is not for costing individual products, as is your traditional income statement, but rather for costing a 'family' of products. So the Lean P&L is far less complex than your traditional income statement. This Lean P&L essentially is a tool that sits outside of your traditional cost system, and has as its sole purpose the planning of any product family's business."

"These Lean P&Ls will be bogus numbers that do not tie to my numbers in the company income statement. Are you trying to undermine my credibility with Jim?"

"You're missing the point here, Nancy," Jim interrupted. "I think Liz is trying to tell you that each product family will create its own actual cost P&L for just that family, in a simpler and less complex format and with fewer categories of cost, for the sole purpose of managing its own small piece of the business. Liz has articulated that this does not involve you, Nancy, and creates no more work for you or your department. And it certainly does not undermine your credibility. You are reporting for the whole business and to external users. The Lean P&L is reporting for just a family and to our internal users."

"Well, then, that brings up another issue. How can I put together a financial package every month for you, Jim, when my department members are sitting, physically, in product families all over the shop floor?"

"Nancy," I interrupted, "as of right now, buyers and engineers and other functional members of the A-team have co-located with the OPS members of their product family in the plant. We have allowed your accounting people to remain as a functional group here in the admin building and work only temporarily with their family until we figure out just how a traditionally centralized function such as accounting will operate in a team-oriented business organization."

"Well, I can tell you right now that I cannot afford to have Carol out of my department and sitting in that family on the shop floor for more than thirty minutes each day. We have too much work to do, and I am already down one head since Marge quit."

"Nancy, most companies today are set up in a centralized structure; however, when we tore down our departments and functional silos and set up product families, we essentially changed our organizational structure to a decentralized business organization. Thus, the question for us to answer is, 'How will the accounting function work if we tear down this glass house and place every accounting person in a product family on the shop floor?' One answer may lie in the fact that we have corporate computer systems which essentially spit out traditional financial statements every month with the push of a button. Other of our sister companies 'close the books' in a day before pushing that button. Then they have a CPA review those computer-generated financials for a few hours, and they're done! So that is one day for the task of external reporting in the traditional functional approach in accounting, and 29 days for the accounting personnel to spend with the family on the shop floor."

"I don't like this!" I could tell that Nancy was about to cry.

"Why don't we just try this, Nancy," I said, "and see how the product family likes it. Already the A-team very much appreciates what Carol has done for her family, and the data she has brought to her family. You must realize, too, Nancy, that this Lean P&L is more than just financial data."

"Let's see how this works, Nancy," Jim interrupted. "If it helps OPS manage the business, we owe Liz the opportunity to show us how this Lean P&L works. We will give this a try, Liz, with this one product family," Jim said to Nancy and me.

After Nancy left, Jim and I continued the conversation. "Nancy's group takes three weeks to close the books," I said to Jim. "I think she purposely elongates this closing process so that she will never be asked to go into the plant. She needs to be challenged, Jim, to get her to close the books in a day as our sister companies are already doing."

"I will encourage her and task her to do just that, Liz," Jim told me.

"I know she has done a lot of reading on Lean accounting, but this is not necessarily something that one learns from books. Also, she needs to understand that the company's focus right now is not on numbers and financial statements but rather on her supporting the physical implementation of Lean in the factory. As Tom has said to me many times, 'Are you going to sit upstairs and move numbers around in spreadsheets, or are you going to come down here and look at how I make the product?' Before, in our first attempt at Lean, we implemented only just-in-time, and JIT is only a Lean tool, not a structural change in the business. The creation of our product families and the moving of furniture and people is a structural change, Jim."

"Liz, I would appreciate you working with her. I want her to become a world-class CFO, as well as interact with other CFOs in our sister companies."

"I think the members of the A-team will be very satisfied at what they can do to understand the cost of their small piece of the business without Nancy and without the assistance of the entire accounting department."

"Perhaps you would allow Nancy to sit in on that Lean P&L training with the A-team, Liz?"

"No, Jim, I do not want her interrupting the group. She would only be challenging everything I say from a purely financial perspective, and that would only confuse members of the A-team. They have no formal financial training, and I like that!"

"Good enough," Jim agreed.

"Are you and Mark playing golf this weekend?" I asked Jim in an effort to end the conversation.

"As a matter of fact, we are, Liz. Why don't you join us?"

"I'm afraid I would only slow you guys down. Besides, my mother is flying in early Saturday morning. She lives more than a thousand miles away and, while we talk by phone for an hour or more every Saturday morning, we rarely have the opportunity to sit face-to-face and chat for hours on end. I am looking forward to her visit."

"Then I'll see you on Monday, Liz. Have a nice weekend!"

"You too, Jim." Then I stopped by Nancy's office, wished her a nice weekend, and told her that we would get together next week, at her convenience, to talk more about this Lean P&L.

8

Mother Comes to Visit

Part of the story of the hospital was told in *My Life and Work*.

> The hospital has nothing to do with the Ford Industries. We own and control the hospital absolutely because we want to carry out in it certain theories which we believe will benefit the public.... The principles that have been given are universal—or, at least, we think so. And we have applied them through all of our industries without finding it necessary to make any changes.... Our principles remain the same....[1]

My sister, who lives close to my mother, had taken Mom to her local airport for a 6:00 a.m. flight. She was arriving at my local airport at a little after 9:00 a.m. on this beautiful, sunny, and warm 80-degree winter Saturday morning. While my mother is up in years, her mind is just as sharp as it was when I was a kid. Even back then my mother always was involved in something, whether it was our school PTA, collecting for the Heart Association, or volunteering at election polling places in our neighborhood. My father had passed away very prematurely many, many years ago. I remember how shocked I was when I got the call at 4:00 a.m. to tell me that my dad was gone. I can only imagine how shocked my mother must have been. He died of a massive heart attack in his sleep. Perhaps my mother knew of his weakness all along, and this was the reason she volunteered for the Heart Association for years.

Very shortly after my dad's death, my mother went to her local hospital and signed up to be a volunteer two full days each week. I believe she felt that she had to do something to keep busy, and keep her mind off her grief or she would die of a broken heart. My sister has been a nurse for 30 years, and so my mother must have heard from conversations with my sister about the good, the bad, and the ugly in the way 'the work' works in a hospital. Thus volunteering at a hospital probably seemed a good choice for her at the time. And I might mention that she has now been with this hospital, as a volunteer, for almost 20 years—the longest service of any volunteer in the history of the hospital.

Mom planned to stay with me for a week to 10 days so she had shipped her clothes to me via UPS. That was cheaper than paying the luggage fee the airlines required, and also made it easier for her to get to ticket counters. She is too small and frail to carry heavy luggage. Today she carried only a small case with her as she left the plane. I am always amazed knowing that the little case she carries is full of meds. How does her doctor know that one of those many medications will not interfere with another and kill her? I get a lecture every time I ask, and so I have learned not to ask.

"Hi, Mom! How was your flight?" I asked as I greeted her just outside her gate.

"The flight was just fine. I always find interesting people on the plane. The man sitting next to me had some interesting stories to tell."

"Well, I am glad it was a good flight. I thought we would stop on the way home at a little place that serves the best clam chowder. Would you like that?"

"Oh, yes! I would love that. I did not eat breakfast as it was much too early when I left your sister's place."

We drove for a little more than an hour, and Mom enjoyed seeing all of the water scenes as we passed over various high and long bridges. The restaurant was not at all crowded at a little after 11 o'clock, and we got a seat at a window, looking out at the harbor. We chatted about things going on back home, and what her neighbors were up to these days. Older people talk a lot about weather events and what they have had to eat lately, so she told me of every luncheon she had been to lately and what they had to eat, and also about the appreciation luncheons she had attended for volunteers at the hospital. Each day they volunteer they receive a free lunch. That is a major social event for all of the volunteers!

When we got to my home, I gave her the opportunity to rest and take a little nap. When she awoke, she started our visit the same way she starts every visit, with the question, "Liz, I do not understand what you do for a living. I understand your sister is a nurse. That is easy enough. But what is it that you do?"

"Mom, I work in manufacturing, and help a manufacturing business implement a business model we call 'Lean.'"

"What in the world is Lean, Liz?"

"It is a business philosophy of eliminating waste rather than cutting cost. Most people think of Toyota when they think of Lean. Toyota is known to be a productivity monster! They can build a car in something like twenty hours, and they build cars at the same rate that their customers are ordering cars so they have no waste in inventory. Just think of all of the money tied up in inventory at the Detroit car manufacturers. I have read that it amounts to billions of dollars! Just think what they could do if they could Lean their businesses, and had all of those billions of dollars in the bank?"

"Why don't the Detroit car manufacturers and other manufacturing companies use this Lean philosophy of business?" Mom asked.

"It is probably because they do not understand the difference between cutting costs and eliminating waste, Mom. Traditional companies are cutting key resources to 'save' cash—doing things like laying off people and closing plants. The Lean businesses, however, are cutting waste—such as not building inventory—to 'create' cash. That is a big difference in business philosophies! Also, the Lean companies are more concerned with productivity improvements, and that implies creating value for their customers. The Lean business is also more concerned with 'processes' than 'transactions.'"

"What do you mean by *value* and *processes*, Liz? I do not know those terms from the hospital business."

"Remember when Daddy died and you had to call the insurance company many times before you got the right person to help you? Well, that insurance company was probably organized in departments, and was oblivious to its customers' needs. If they were organized around the customer, you would have had to make just one call. Also, you had a direct flight today but, if you had had a connecting flight, the airlines would have routed you through whatever hub was most convenient for them— perhaps Chicago! You would not have liked that! Companies today seem not to be concerned with creating value for their customers. But let's see if we can relate the meaning of creating value to your hospital."

"That would be great, Liz!"

"One way we have to help us create value for the customer—which is the patient in your case—is to create a matrix. In manufacturing we might place all products we sell along one side of the matrix, and all machines or processes across the top of the matrix. In the case of your hospital, we might place all of the room numbers (we don't want to use patients' names) along one side of the matrix, and all of the meds in the hospital pharmacy across the top of the matrix. In manufacturing, we would place an x in every box for every operation that a product passes through, then group products that run through like processes into families. In your hospital, we would put an x in every box for every med that every patient takes, and group patients together in families based on the fact that certain groups of patients take the same meds and thus have many other commonalities in illnesses and thus treatments. Families at Toyota might be the Lexus, the Corolla, the Prius, etc. I would imagine that families in your hospital might be Cardiac, Oncology, Pediatrics, and so forth. Now, once we have the various families established, we move the furniture so that the workers have everything they need in one place."

"I see. But what do you mean when you say that a Lean business is more concerned with processes?"

"Processes in your hospital might include admitting, discharging, record keeping, and various medical procedures such as MRI. These processes all add time and cost to a business. Let's take that cup of coffee we are both enjoying. I made decaf for both of us. Would you think that decaf coffee would cost more or less than regular coffee?"

"Well, I know it costs more, but I think it should cost less because there is less in it. They took the caffeine out of it!"

"Yes, but they had to do another process to eliminate that caffeine. And additional processes cost additional money, Mom. Henry Ford would have lined up all of his processes in making his Model T, and taken out the processes that did not add value, or were waste, for the customer."

"Oh, funny you should mention Henry Ford. Do you remember the Model T your uncle Larry had?"

"I don't think so, Mom, but the point I want to make here is that Henry Ford, not Toyota, was the person who actually came up with this idea of Lean—or what Henry Ford probably called 'common sense.' So you see that Lean is not a fad, as some would have us believe, but it is a business philosophy that has been around for a long time—almost a hundred years!"

"If it has been around for such a long time, Liz, why don't more people utilize it?"

"That is probably because they do not understand the principles of Lean, Mom."

"And what are those principles?"

"Well, the first principle is to create value for the customer, and then the second principle is to 'map' that 'value stream.' We have already talked about value. The value stream in your case would be the movement of the patient through the hospital. If you created a map of a patient moving through the hospital for his various treatments, would that map look like a plate of spaghetti? The third principle is to create 'flow'—to line up all the operations in a product family in the order in which the operations occur. I have a really funny example of this. Remember when I was up at the university for a week for some Lean training?"

"Yes, I do. That was the time that nice friend of yours in your class brought you home to spend the weekend with me."

"Yes, that was the time. Each day we had 45 minutes to go down to the cafeteria to get some lunch. There were a lot of classes in that very large engineering classroom building, and so there seemed to be hundreds of students waiting in line every day to get lunch. We would stand in line for forty minutes, and then have only five minutes to eat our lunch. It was very stressful. So one day a couple of students from my class suggested to the cafeteria manager that they Lean this process. The students suggested to the cafeteria manager that they form two lines inside and outside the cafeteria: one for hot lunches, and one for cold lunches. The cafeteria personnel took the advice and two lines were formed, cutting in half the amount of time we had to spend standing in line. But there was a hiccup."

"Really? What was that?"

"Once inside the cafeteria, the people who were in the hot lunch line were crossing over to the cold line to get a pat of butter for their hot roll, and people in the cold lunch line were crossing over to get a cup of hot soup to go with their cold sandwich or salad. What a mess!"

"That is funny, Liz. Perhaps we could talk more about this Lean philosophy of business in terms of my hospital after dinner this evening?"

"Sure, Mom. Just tell me what is on your mind, and we can knock it around. In the mean time, I have a nice piece of grouper to put on the grill for us this evening. Would you like to sit out on the lanai for a while before dinner?"

My mother actually went out to the pool for a while, which amazed me. She had always had a fear of water since the time her older brothers played a prank on her as a child and held her under water. When she came back in, she was all excited! She had met another older couple who were from a town near where she grew up. Suddenly this water community, which she never liked, was becoming something of much interest to her.

Mom was tired after dinner, and so she retired early. It had been a full day for her. On Sunday morning we went to a local church service, and then came back home to the island for a meal and more talk.

"May I ask you some more specific questions about how this Lean philosophy of business might apply to my hospital?"

"Sure, Mom. I don't know if you know this, but I did my master's thesis on improving cash flow in a hospital using the principles of Lean. I love talking about this! I just do not want to bore you."

"You are not boring me, Liz. You know I like learning new things. Just recently the hospital put me at the main desk all day. My job is to use the computer to look up the room numbers of patients when their friends and family come in to visit them, and then tell the visitors where their loved ones are located and how to get there. I loved learning to use a computer! I would even like to have one at home now."

"Wow! So what is on your mind, Mom, and what is going on at your hospital now that has you so concerned?"

"Well, you know that the head of the volunteers has become a good friend of mine. She is young and very smart, and she shares with me some of the concerns of the hospital execs."

"So what is her concern today?"

"She tells me that we need twice as many nurses as we have just to keep our operations going, but we cannot afford to hire more nurses. We may have to shut down the hospital for lack of resources. What would you do, Liz?"

"Hmm. I would probably do a value stream map to see where the waste is. Waste can be not just in inventory, as with the car companies, but also in the form of waiting time and excess motion or transportation. Something like a service is not value-added if we have to wait a long time for it, or travel a great distance for it. Perhaps you should ask your friend who is head of the volunteers if you might follow just one nurse for one eight-hour shift, and 'map' where she goes, how long it takes her to get there, how long she stays, and what she does there. I believe you will find that a nurse's job is a blizzard of activities that span over the entire hospital instead of confined to just the space of her few patients. I bet you will find that this one nurse spends 80 percent of her time just looking for what she needs to do her job and only 20 percent of her time fulfilling her responsibility to her patients. If that is true, then you do not need to hire more nurses."

"So, if I understand you correctly, you are saying that we want to look at our hospital business, not in terms of departments, but rather in terms of 'families' and what you call 'value streams'?"

"That is correct, Mom. Instead of an MRI department, for example, you need to get rid of that one huge monument and get smaller MRI machines to place in each family that needs one. You do not need monuments. You need toasters! You want to bring the mountain to the patient, not the patient to the mountain. When you shrink that value stream you created with your value stream map and take out those things that are waste and interfere with flow and do not add value, you are practicing cost containment! That consulting firm I was with years ago did a study that showed that it is possible to take 85 percent out of the lead time of a health care business utilizing the Lean philosophy of business. Don't quote me on that number as it has been a while since they conducted that study, but an 85 percent reduction in lead time can translate to substantial cost savings."

"Yes, that would be a spectacular feat, Liz. I cannot help but think, too, of the time your daughter was hit by a car and almost died. Have you ever thought of value stream mapping the accident that almost cost my granddaughter her life?"

"Yes, as a matter of fact, a friend of mine was in a similar traumatic accident many years ago. He also, like you, asked me about 'Leaning' the trauma business, and his case illustrates yet a third principle of Lean, which we call 'pulling' from the customer. But that is another story for another visit. We need to clean up this dinner, and get relaxed before bedtime. I do not want you having medical nightmares!"

"I have really enjoyed our discussion, Liz."

"Me, too, Mom. I am glad that you are here."

REFERENCE

1. Ford, Henry, and Samuel Crowther. 2003. *Today and Tomorrow.* New York: Productivity Press, pages 172–224.

9

The CFO and CEO Push Back on Lean Cost Management

The day after a Category 3 hurricane a few years ago, my neighbor received a call from his landscaper. She asked him, "What would you like me to do with your yard this month?" and he responded, "Probably nothing if there is no house there anymore!" The moral of the story is that, if you do not have a cost management system, there will be no need for a financial accounting system as the business will cease to exist.

Mom was up with me at 5:00 a.m. I cannot remember a day when both of my parents were not up at 5:00 a.m. She had insisted last evening that I not take any time off work while she visits with me. She said she had plenty of things to read, and she had made plans to meet her new friends at the pool again today. She also said that my location was perfect for some nice walks during the day. As Mom made coffee, I got dressed for work. I planned to stop in to see Jim and Mark first thing this morning. Jim had asked about the difference between cost accounting and cost management—for the third time. I still have this feeling deep down inside that he is going to cave in again. I want to talk with him with Mark in the room so that Mark can help me read between the lines. Then I wanted to talk with Nancy, perhaps very casually over lunch. Of course, the main event of the day was to meet with the A-team to create the Lean P&L.

Mom waved and laughed as I drove off in my golf cart, headed for the dock to catch the first boat off the island. Once I got to the plant, I headed straight for Jim's office. I had left Mark a message, asking him to be there also. Nothing dramatic was happening—just a sinking feeling that Jim's commitment might falter in the near future.

"Hi guys! Who won the golf game?" I asked both Jim and Mark as I approached Jim's office.

"I let Jim win," Mark replied.

"So," Jim said, "is today the day you meet with the A-team to discuss the Lean P&L, Liz? I would like to know more about this myself. I still do not understand why we need this. Why can't we use the same financial system we have now?"

"Our present financial system is a standard cost accounting system, Jim. It is all about transactions, debits and credits, and standards. It does nothing to support OPS. A Lean cost management system, on the other hand, is all about planning. We plan our revenues. We should plan our profits and, if we really did plan both, we would be planning our costs. Costs should not just happen! We need to know how OPS methods—the physical operations—affect costs. You have already seen how lining up machines in the order in which the operations occur reduces time and cost."

Mark chimed in to help me. "We need to trace costs, Jim, and we cannot do that with our old departmental organization without the use of huge IT systems. Everything in the current financial system is so complex and so outdated and so time consuming. Can you make a good decision today with the latest financial statements that came out over six weeks ago?"

"And, besides," I said, "in the past we were acting like Lone Rangers or hammers looking for nails, executing only point improvements. That is terribly myopic. We were not even looking at the whole business. We were not thinking *systems*! I cannot emphasize that enough, Jim."

"I am afraid I still do not really get it, Liz. Our friends on the other side of town are doing Lean, and doing well, and the CEO of that business tells me not to even think about Lean accounting."

"Jim, Lean will actually automatically re-create our financial system for us! Our financial accounting system—a standard cost absorption system—tells us to keep all of our people and all of our machines busy making product 24 hours a day, seven days a week for the purpose of minimizing unit overhead. But Lean—if you understand the principles of value, flow, and especially pull—tells us to only 'make one' when the customer 'takes one.' The old accounting system is encouraging bad behavior, Jim. It is encouraging everyone to build inventory!"

"I see. You do have a point, Liz."

"You can still use this standard cost system to value inventory on the balance sheet. That was what the system was designed for originally. But we cannot use it to manage the business, Jim. It is creating all the wrong behavior, and will sink this ship a second time!"

"But we did move all of the furniture for these product families, as you and Mark requested. Isn't that enough?"

"Work habits do not change when you just move furniture, Jim. To be successful in Lean we must change behavior, and to change behavior we must change our standard financial cost system to a Lean cost management system. Let's look at some of the many things about our current standard cost accounting system that are not helping us. First, the financials are too complex and in a language that OPS people do not understand. I would challenge even the accountants to explain some of the many standard cost variance calculations. The cost system should serve the OPS people—not the financial people. If OPS people cannot understand these financial reports, how can they improve them? Right now, after you berate OPS people for not 'making the numbers,' they just toss these traditional financials in the round file.

"Second, this financial system gives you and others reports that come out about six weeks after the event. Do you actually make decisions based on info that is six weeks old? Would you buy or sell stocks based on prices that were six weeks old?

"Third, this standard cost system is challenging us to 'beat' the standard as our goal this year. When was the last time that labor standards were updated? We laid off industrial engineers years ago. As for material standards, I can set a standard that I know I can beat. So here we are in the game-playing mode again! These long meetings to discuss and chase variances to standard are a waste of time, Jim. Our time would be better spent talking with our customers.

"Then, most important of all, this standard cost system does not support any continuous improvement. Again, we get a data dump of debits and credits that are weeks or months old. What kind of information is that?"

"I would have to agree with her, Jim," Mark pitched in. "OPS people are discouraged, and have given up. If something does not change, they will just go through the motions of doing whatever it is they believe management wants them to do to keep their job. They will not be concerned with making the business better. They don't know how to do that! Just last month someone in accounting presented some monthly financial info to my group. When she finished scolding us, a person in my group asked what made up the metric 'cost per case' so that we could improve it. This finance person had no idea—no suggestions for how we could improve. And worse, she ended her presentation with, 'Better luck next month!' and she laughed. The complexity and mystery in this business just continues to build and build."

"I think I am beginning to see your point, Liz and Mark. I agree that we do not want a financial system that allows people to manipulate numbers."

"And we do not want a financial cost system that is all numbers and ignores the OPS people it was meant to serve. Right now we have a cost system that is dependent on a batch style of manufacturing. That does not fit with our Lean manufacturing, which is a one-piece flow model. We have made changes in the design of our operations. Now we need to make a change in the cost system which supports our new Lean manufacturing design."

"Well, please answer just one more question for me, Liz. How can my friend across town be so successful with Lean when he has a traditional accounting system? If all that you say is true, how can this be?"

"Your friend across town has been working Lean for six years, Jim. He has his inventory turns above twenty-four turns per year. That tells me he has no inventory. If one has no inventory, then there is no difference between absorption cost accounting and a variable or direct cost system such as the one I am proposing for our Lean business. We need this direct Lean cost management system until we get our turns from 2 to 24 times per year, Jim."

"OK, I surrender. I do think I get it. You did tell me that the first year or so we might show a loss due to the fact that we are trying to use up our existing stash of inventory."

"That's good, Jim. I know it is very difficult to change the way one thinks after years of being taught a certain way at our universities and also having worked in a certain way for many, many years in old world business. This is why they say that change is difficult."

"I really would like to hear what you have to say to the A-team this afternoon, but I know you do not want me or Nancy to be there. Might Mark be there and tape it for us?"

"I do not have a problem with this if Mark does not. Or perhaps a better idea would be for me to just repeat this session with the entire Steering Team so that we are all on the same page. I do believe that might be a better idea than taping what questions these A-team members may have. That would surely violate their trust in me and you. We could do this after hours with the entire Steering Team, and decide on our Lean metrics at the same meeting."

"Sounds good to me," Jim and Mark said in unison.

I stopped by Nancy's office, and invited her to join me somewhere way off campus for lunch. She agreed, and we met in the parking lot a while later. We decided on a little Thai place where we could get in and out quickly and not be seen by others in the business who prefer the sandwich and pizza shops.

"I met with Jim and Mark this morning. Jim was pushing back on this Lean cost management system. I want you to do the same thing, Nancy. Push back on me, and ask questions. I do think you get it. You're just trying too hard to tie this Lean cost management system to your traditional standard cost accounting system."

"Well," Nancy said, "should the two systems not tie in some way?"

"Not really. The two systems serve two totally different purposes, Nancy. Your financial system is for external reporting, and to value inventory and COGS. It is a snapshot in time of the whole business. The Lean cost system is for OPS people, to assist them in managing their piece of the business every day, making decisions as to what products the customer wants, and how to cost and price those products. Its purpose is to help plan a small piece of the business."

"I see. But you have spoken so much lately, Liz, about systems and linking everything together to form one large system. Why do you not want to link these two cost systems?"

"They will never link, Nancy, because they are two totally different systems serving two totally different purposes. And they could never link as the traditional system is in standard cost while the Lean system is in actual dollars. Your system shows variances to standards, and the Lean system shows variances of actual costs to targets (which we will develop soon) and trends. Your traditional system and the Lean system serve two totally different customers, Nancy. Trying to reconcile an actual cost system (with variances) would be considered a waste and non-value added."

"But we need all of the information in the financial system, Liz."

"People outside of the business need the information in the financial system, Nancy. But do OPS people need all of the info in the traditional system? I think not. OPS people do not need an elaborate complex system such as the traditional standard cost system. You need a lot of IT horsepower to run this financial system, and a lot of people to update it. The purpose here, from the viewpoint of operations, is not to automate cost but to eradicate cost. Automating this does not make it any easier to read and understand. They are looking for something simple, something in a language that they can understand, and something that goes together fast so that they can have this any day they need it. This Lean system will sit outside of your formal system, Nancy, for the sole purpose of managing and planning a piece of the business. Think of it as a *tool* for operations."

"I still do not really see the purpose in this Lean cost management system, Liz."

"Perhaps if you look at it this way, Nancy: it is meant to serve only OPS. The costs in it are actual costs, not standards. Think of it as a checkbook for the people in the product family. It is something they can understand, relate to, act upon, and improve upon. And, most important, every cost in the Lean P&L is a direct cost. And by that I mean that every cost in that Lean P&L is a cost that the family can directly control and for which the family can be held accountable. This creates value for them! After all, the purpose of any cost management system is to serve operations. Look at this visual I brought with me. I showed this to Helen, and should have shown it to all of the Steering Team. Perhaps this will help you understand." (See Figure 9.1.)

"OK, this does help, but what do you mean when you say that 95 percent of costs are direct costs?"

Our New LEAN Structure...

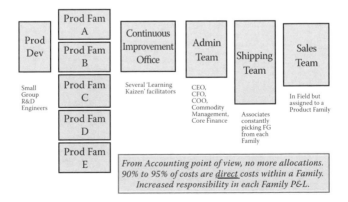

FIGURE 9.1
Physical structure of our new Lean organization.

"The only costs that are not direct costs are outside the product families in the diagram. Thus you can assume that material, all labor (including what we accountants would call indirect labor), outside vendor services, all supplies, machine depreciation and scrap, utilities, etc., are all direct costs within each family. The family controls these costs. We do not want to allocate costs in the Lean P&L. Allocations are arbitrary and distort product cost, plus the family members have no control over improving cost allocations. Thus the 'a-word' is a 'no-no' in our Lean families. Does this make more sense now?"

"Not really, Liz. Why do we want this type of setup?"

"We want to make things simple in Lean, Nancy. We do not want to aggregate indirect costs in large pools and allocate them to products or cost centers. This is not transparent! People in OPS do not like to have costs shoved down their throats, and then be told to reduce those costs. You cannot hold people accountable for those costs they have no control over, Nancy. In Lean cost management, we are essentially tracing costs within a product family instead of trying to track them all over the business and allocate them to a single product with a huge computer system."

"I think I am beginning to see your point, Liz. When we have our daylong meeting to discuss cost variances, everyone is sent off to investigate these unfavorable variances to standard for days and then report back. If we are instead managing costs in the product families, which are smaller entities, it will be easier to see what is causing the costs. Am I right?"

"Yes, you are, Nancy. Remember when we talked about tying together QCD—quality, cost, and delivery? Each product family, unlike a cost center, is now dedicated to a specific group of products and a few specific customers, and the family has control of all costs—and revenues! In such a setup they can deliver on not just cost but revenue, quality, and speed to market as well."

"I believe this is all representative of what they call the 'thinking processes' in Lean business books," Nancy remarked.

"You have that right, Nancy! And Jim just said the same thing this morning. I think we had better get back to the plant. I enjoyed hearing your perspective, Nancy. Just remember that these two systems each have their own purpose, and their own customer. They are apples and oranges, but both necessary. In time, when we have inventory turns in excess of 24 and basically have no inventory, then the two cost systems—one absorption and one variable—will be similar."

I had just enough time to call home before heading over to the A-team to introduce the Lean P&L. Mom was not answering the phone. I hoped she was OK. I should have left an emergency number for her. Now I will worry all afternoon.

10

How Will the Lean P&L Work?

In almost any business attempting Lean today, there is a complete disconnect between the accounting system and Lean operations methodology. We are using a standard cost financial system that does nothing to help operations people manage the business day to day. We accountants chastise operations people for not "making the numbers," but we don't give them any suggestions on how to improve the business. Why? Because our standard cost system does not give even the accountant such information for a Lean entity, and because our traditional standard cost system focuses only on the balance sheet and not on the shop floor. So not only do we need an alternative cost system to reflect the physical and operational design changes that are a result of our Lean initiatives in the plant, but we also need to gear the financial mumbo-jumbo to something the average working man in the shop can understand and impact.

The A-team was waiting for me, and they seemed excited. Things had been going well with the setup of the family, and everyone was getting along well. Everything was "gelling."

"Hi, everyone! How's it going?"

"It's going great!" said Harry, a shop member of the A-team. "The company has given us the responsibility for a group of products and a group of customers. This is fantastic! For the first time ever we truly are empowered to make a difference!"

"Yes, and we see so many ways to improve our piece of the business," Bob said. "Mindy has been leading the egg hunt for inventory, and we have all of that organized downstairs here in this building where all of our manufacturing takes place. Now we can get rid of all of those trailers that we had leased just to store all of this stuff."

"And better yet, we may be able to get rid of that old building across town where our value stream begins. That entire building—both floors—is used solely for the purpose of receiving our raw material, and then checking the quality of the raw material. We need to do what you and Mark did in packaging and shipping, Liz, and make the supplier responsible for quality. Then we could get rid of that building, in addition to all of those trailers or, if we own it, we could sell or rent it to another business in town."

"And there is more news, Liz!" said another shop worker on the A-team. "During our egg hunt for inventory, we discovered so many assets that we no longer use—things like office furniture, old machines, and old tools. We remember our Learning Kaizen involving 5S. We were thinking of placing an ad in the local paper for a 'yard sale' to get some money for this stuff rather than just throw it away. Carol says that would increase our ROA or return on assets, and that is a good thing. What do you think, Liz?"

"I think you folks are on a roll! You are looking at your value stream, taking out the waste, and that includes taking out processes and assets that do not add value. You really do get this!"

"This is fun, Liz! The company truly has given us responsibility, and we are taking it seriously."

"Well, I have come to give you even more responsibility—for the cost of your family. Are you ready for that?"

"We sure are. Bring it on, Liz!"

"Let's sit over here for a while. I have some questions for all of you, and we will need a board on which to write."

"OK, let's all go into our break room."

"What costs do you folks consider important in order to have control over your piece of the business?" I directed my question to everyone in the room.

"Definitely material," someone said, and I wrote "Material" at the top of what would be the list of direct costs in the A-team's Lean P&L.

"We also care about our labor costs," someone else added.

"OK, and I believe we want to include all labor in this line item," I said. "So here," I said as I wrote "Labor" on the board, "we will record not just manufacturing labor but also supervisory labor plus all of the benefits the company pays to both groups of workers. What else do you want to track in the way of costs?"

"Well, I think we need to track supplies we purchase—things like oil for the machines, and special tools, and stuff like that."

"Good point!" and I wrote "Supplies" on the list of costs that they would be held accountable for. "What else?"

"We often have outside vendors that perform certain services for us," someone remarked.

"OK, so I will add 'Services' to the list of costs for your family."

"What about material handlers? Where do the costs go for these people?" someone asked.

"We could place them in the 'Labor' line item. They are what accounting would call 'indirect' labor; however, for us in the family, 'labor' is labor. All costs we list here are 'direct' costs because you in the family are directly responsible for these costs. This Lean P&L will serve you, the family, not the accounting department."

"What about depreciation of machines?" Carol asked. "I know that is usually something that only the accounting group is concerned with, but I think our family should be concerned. While it is not a cash expense, depreciation expense is a way of setting aside money for the replacement of the machine at the time when it will supposedly have expired in it usefulness."

"Good point, Carol. I agree that depreciation of machines should be in your list of costs for which you are responsible. Then what about scrap? Does anyone care about the cost of material you scrap?" I asked.

"We should be concerned," someone shouted. "We throw out a lot of raw material so, until we make the vendor responsible for quality, I think we need a line item for 'Scrap.' The cost of scrap is most justified."

"OK, then what about electricity, heat, and any sort of 'Utilities' expense. Do you think you should be responsible for these costs in your family?"

"Sure, we should! If some idiot in our family goes home late and leaves all of the lights on and the heat or A/C turned up, we are responsible for that. Also, being responsible for power would encourage us to get our product made in as few shifts as possible and to keep our machines running efficiently so as to use less power. But how would we do this?"

"We could have special meters installed to track the power your A-team is using, both in machinery and in climate control."

"What about the cost of inventory?" someone asked.

"In order to keep this Lean P&L simple, I think we will keep the increase or decrease in inventory (both material content and labor content) out of the direct material and labor costs of the P&L, and place the increase and decrease in inventory below the total cost line as a lump sum, then also have inventory turns as one of our Lean metrics. Thus you will be held responsible for the dollar increase or decrease in inventory in each period, which can be calculated with a simple cycle count, as well as the metric of inventory turns (which is cost of sales divided by average inventory) for your family. The same would be true for the cost of the floor space you 'rent' from the corporation. The accountants would have that as an occupancy cost in your P&L, but that is an allocated cost and we do not want any allocated costs in your Lean P&L. Thus we will keep this occupancy cost out of your Lean P&L, and place your responsibility for shrinking your occupied floor space in your Lean metrics for now."

"What about all of those other corporate costs that accounting always allocates to our product line?" someone asked.

"Those corporate, or admin, costs are not your responsibility. You cannot be held accountable for what the corporation decides to pay the execs or the legal beagles or the R&D engineers, as someone mentioned earlier. They are outside of your family. Look at this visual (Figure 10.1). Here you see a schematic of our old business layout. Some call this the smokestack organization. The costs of the people in each of these smokestacks are allocated to your operations department in our old traditional organization and in our traditional P&L.

"Then look at this visual of the structure of the new Lean organization (Figure 9.1). Here you see those costs that are outside of your family, and not your responsibility. And that implies that the costs of these people will not be allocated to your family in the Lean P&L. Some of those costs are

Our OLD Structure....

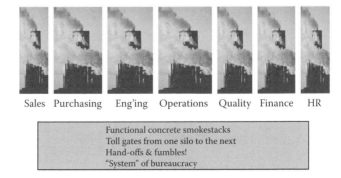

Sales Purchasing Eng'ing Operations Quality Finance HR

Functional concrete smokestacks
Toll gates from one silo to the next
Hand-offs & fumbles!
"System" of bureaucracy

FIGURE 10.1
Physical structure of our old functional departmental organization.

the costs of the R&D engineers who develop new products, the Learning Kaizen facilitators, those people who will stay in what I would call a centralized area of the business (i.e., corporate finance, commodity management, and our executive group) and the shipping people. The corporation is responsible for these costs. They are not in your family. Corporate will track these costs, and be accountable for them."

"What about the sales team?" someone asked.

"You do understand that you have a person from sales assigned to your family, even though he is in the field and his wages and benefits and his bonuses, if any, are the responsibility of corporate. Also, within your family footprint, you have an engineer, a person from quality, a person from HR, a person from procurement, and a person from accounting, as well as many OPS people. The wages of all OPS people who are within your family are your responsibility. Likewise the wages of Carol here, and the other salaried members of your family, are also the responsibility of the family. While these salaried members of your family work in your family, their wages are fixed for the year. They are not paid overtime. They do, however, now report directly, that is, 'solid line,' to the manager of your product family and indirectly, or 'dotted line,' to management. But most important of all, they reside physically inside your family, and you are responsible for the costs of these people."

"I think I understand," Carol added. "So the family will have a big say in whether or not these salaried persons did a good job in the family at the end of the year, and the family will have a say in what or how much of a raise these persons receive for the next year. But perhaps we would want benefits as a separate line item under the line item of labor, and perhaps, too, we would want separate line items for hourly labor and salaried labor."

"Good point, Carol. We do want to be able to trend just hourly wages as opposed to trending all monies paid to all associates in the family, including the salaried folks."

"I like this!" someone from the family exclaimed. "We really are being held responsible only for those costs over which we have control."

"So to summarize all that we have talked about here, Liz, we can conclude that our new Lean P&L contains only these cost line items: material, shop labor and salaried labor (and associated benefits for all), supplies, services, machine depreciation, scrap, and utilities. Then, below the total costs, there will be a line item for the increase or decrease in inventory. Is that correct?"

"Yes, it is. And please note that your Lean family is a profit center—not a cost center as it was in the old organization. You will be responsible for your revenues as well. You are also responsible, of course, for the square footage of the space you occupy, but for now we will track the amount of floor space you utilize in your Lean metrics which we will introduce soon."

"This is truly unlike the traditional P&L I have seen the accounting folks distribute at month-end. That P&L is at least four pages long, and includes everything but the kitchen sink! I would call our simple Lean P&L a checkbook P&L as it is just like the checkbook for my family at home. It has only a few line items, and it is in a language I can understand. Best of all, it allows us to manage and plan our own piece of the business."

"I might suggest we call it our 'Action P&L' because we can truly act upon or influence every single one of the line items in this scorecard," someone else commented.

"It is like a scorecard, isn't it? It really is not a formal accounting statement. It is not for accounting! It is just for those of us in the family, to help us manage our costs day to day and plan our small piece of the entire business."

"It is also most unlike one of the formal accounting statements in that you are tracking your costs in actual dollars, not standard cost. For example, material costs in your Lean P&L are what you actually bought and used for the period. The same with labor costs, supplies, etc. You are tracking only what you actually spent in the month, quarter, or whatever period you choose to manage. No more trying to explain variances of actual costs to standard cost, and accruing costs. You will be looking at your actual costs this month compared to those actual costs last month or for the same period last year, and then to targets, which we will develop for the year-end. Trends become more important than variances to standards, and will be something you will see if you graph these actual costs—which I would encourage you to do. Mark can give anyone you designate some training in how to do graphs in Excel. For that matter, all of your Lean P&L costs can be tracked in Excel, and then easily converted to graphs for visuals."

"This is great, Liz. I am really glad we started this Lean transformation a second time, and I am glad we are doing it correctly this time."

"The only thing different this time around—and it is different—is that we are focusing on a physical transformation of the organization rather than just attempting to implement tools such as JIT and kaizen. These families—and there will be more—will link operations and cost, and create our new Lean cost management system."

"Thanks, Liz."

"You are welcome. I am delighted that you like this. It is fun, isn't it? Let Mark know if you need any help with anything, or get back to me. See you around!"

I was pleased at how well that went, and so I headed home to check on my mother. I hope she has not had a stroke or something worse!

As I pulled the golf cart up to the house, I saw her sunning her legs out by the pool, reading one of Henry Ford's books from my bookcase. Oh my goodness! Is she serious? I believe I have created a monster!

"Hi, Mom! How was your day? I called home at noon, but you did not answer the phone. I was worried about you."

"Remember, I told you that I was going to meet your neighbors out at the BBQ area at noon? We had a little picnic. It was delightful! She and her husband have a big farm just like the one I grew up on, and in the same area! We had a wonderful talk."

"I am glad you had a good day. Now, what is that book you are reading?"

"Oh, this? I just picked it up off your bookcase. I got to thinking about what you said when you described processes. You said that I need to look at all the processes and determine which ones do not add value. Well, I was thinking about the time I worked in Admitting. That is a process, right? A new patient may spend two to three hours going through Admitting when they first come into the hospital. They are nervous, and they do not feel well, and they probably got up really early that morning. All of this adds a lot of stress to the whole experience."

"What did you have in mind, Mom?"

"What if we 'admitted' new patients via phone or mail before they ever come to the hospital? Then they could go directly to their room in their 'family,' and be greeted by the team of doctors and nurses who will attend to them. Is that not what you said, too—that we should bring the mountain to the patient, and not the patient to the mountain? I think this would be far more comfortable an experience for the patient, and it would eliminate waste in the form of a physical 'step' in the line of 'processes,' would it not?"

"Yes, it probably would, Mom. You are amazing! You have this Lean stuff in your blood. However, what I intended to do in our discussions was to give you a methodology for discovering the root cause of the problem in your hospital. Then it is up to you and your hospital associates to determine a solution to the problem that meets the needs of your hospital. I cannot do that for you. The solution to any problem is not some recipe one reads in a book or the solution that some other business implemented, but rather it has to be discovered by trial and error by those of you in your hospital. The solution is never predetermined by copying someone else's solution. You and your hospital friends will make mistakes, but you will learn something from those mistakes."

"So what's for dinner? I could get things started while you change clothes and check your mail."

"I made a German potato salad before you ever got here. It is in the refrigerator. And I brought home with me another piece of fresh fish. Perhaps you could choose a veggie you like and get that started."

"Will do, and then perhaps we can take a buggy ride after dinner. I have not seen much of the island yet this trip."

"That sounds great! And we could even stop by Coconuts and get a soft ice cream for dessert!"

11

Which Comes First: The Chicken or the Egg?

Structure drives behavior which creates a culture.

I have my mother in the car with me this morning. I am dropping her off at the home of one of her northern neighbors who just bought a winter home in this area. She will spend the day visiting with them. I will spend most of the day with the Steering Team, discussing our Lean progress to date as well as the Lean cost management system.

Jim, Nancy, Helen, Mark, Tom, and Greg were all gathered in a small conference room off Jim's office when I arrived.

"Good morning! How is everyone this morning?" I asked.

"We are good, Liz, and eager to hear an update on the meeting you had with the A-team yesterday. How did they receive the Lean P&L? And will you be discussing this with the committee today? We sure hope so! We are all eager to know more about this Lean cost management system. None of us knows much about Lean accounting."

"I am eager to present all of the above to all of you this morning, and answer any questions you may have. First, let me precede what I have to say with the news that the A-team is progressing well. They feel empowered, motivated, and are all very busy working well together in their new product family. Mark-san did a great job in designing the layout of the A-team family, as well as choosing the members for the team."

"Thanks for the compliment, Liz, but I was just creating the value stream map and flow for the family. It is what I have been trained to do in Lean implementations."

"Well, with value and flow in place, now it is time for the financial piece of Lean. As some of you may not know, I have made it a habit to attend Lean leadership conferences whenever possible. At one such conference several years ago, there were in attendance many middle managers from very large, well-known businesses engaged in Lean. It was a fabulous opportunity and privilege to network with these people from all over the world for two days. At any of these conferences, I always ask the same question of the Lean experts: 'Why do some companies have such ugly cultures?' I never get an answer. But, at this particular conference, there was a table of five middle managers from a company very well known for its commitment to Lean for decades. One of these gentlemen turned around to face me, and placed a small paper in front of me. He had written down what was, to him, the answer to my question. And his answer to the question of why some businesses have such ugly cultures read as follows:

STRUCTURE DRIVES BEHAVIOR WHICH CREATES A CULTURE

"I want everyone to think about this. When we all started Lean the first time, we were focused on JIT—a Lean tool. We did not change how the business organization was physically structured. That is to say, we did not change the design of operations or how "the work" works. We never got to the first three principles of Lean: one, create value, that is, the creation of product families to replace departments and functional silos to better serve the customer; two, map the value stream, taking out waste and non-value-added activities; and three, create flow, that is, lining up the machines end-to-end in the order in which the processes or operations occur, eliminating batches and queues. Both Jim and I have expressed a desire to change people's behavior in order to be successful at Lean this time. We do that, for starters, by changing the way the business is physically structured. We started this time by moving furniture—and people. This new structure drives new behavior, which will then create a new Lean culture. But then we will have to ensure that this new culture or new behavior sticks or stays in place. And we do that with a Lean cost management system.

"I have brought two visuals with me. I want everyone to look at the first diagram of functional smokestacks. (See Figure 11.1). This was the physical structure of our business when we began our first Lean initiative years ago. We were organized in functional silos or departments. We had a lot of

Our OLD Structure....

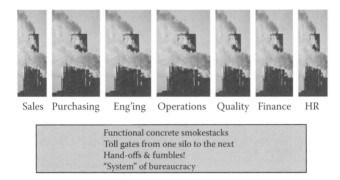

| Sales | Purchasing | Eng'ing | Operations | Quality | Finance | HR |

Functional concrete smokestacks
Toll gates from one silo to the next
Hand-offs & fumbles!
"System" of bureaucracy

FIGURE 11.1
Physical structure of our old functional departmental organization.

goal incongruence. And the only 'system' we had in this business structure was one of bureaucracy. We had attempted to improve or Lean the business with what I call 'drive-by kaizens' and JIT, but we just became more myopic and nearly 'jitted' ourselves to death, ignoring the Lean principles as well as the physical structure and design of operations in the company.

"Now please give your attention to the second diagram—the one with five product families. (See Figure 11.2). This is the proposed physical structure of our organization in our most recent second attempt at Lean. Members of every functional smokestack have been physically taken out of the smokestacks or silos, and placed in a family. The only people sitting outside of these families are our R&D folks, our commodity management team, and a core financial group whose job it is to produce traditional financial statements for external corporate and supervisory bodies and investors, our kaizen facilitators, our sales people, our shipping people, and our executive team. Thus, I would say that 90 to 95 percent of our people, and thus our costs, are now within the product families. Would you agree?"

"Yes, I do," said Helen, "but surely you are not implying that merely moving furniture changes behavior!"

"Moving furniture and people is a step in the right direction, but you are correct, Helen. It is not the end all. However, it will help with communication among 'functions' and allow us to put this Lean philosophy of business into practice one step, or one product family, at a time. I think

Our New LEAN Structure...

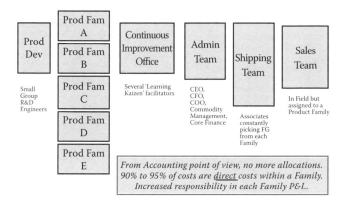

FIGURE 11.2
Physical structure of our new Lean organization.

everyone would agree that a pilot is better than attempting to execute this Lean concept into the whole business at once."

"You are talking about Lean 'cost management.' But how is this different from what we know as 'cost accounting'?" Helen asked.

"Jim has asked many times about the difference between cost accounting and cost management. Cost accounting is what we have now. It is an elaborate process of collecting indirect costs, which are far more than 50 percent of the cost of our business and everything except material and direct labor, and then arbitrarily assigning them to individual products via a complex allocation methodology and IT system. As this legacy system was developed decades ago, no one today truly understands how it works and so everyone is afraid to change it for fear it may 'break,' and no one understands it well enough to 'fix' it. We use this cost accounting system to value inventory on the balance sheet, but this system is not good for continuous improvement or cost reduction and should not be used to manage the business. Henry Ford once said that cost accounting, as opposed to cost management, tells us that something is wrong, but not what is wrong. Lean cost management, on the other hand, is the process of making all costs direct costs and tracing them to families of products for the sole purpose of managing or planning a piece of the business. Cost accounting is for accounting; cost management is for operations. Cost accounting might be considered by some to be a sub-ledger of the financial G/L. But a cost management system is not a formal accounting system. It is far more than numbers, and what numbers are there are actual costs—not standard costs."

"But we do have a cost management system right now, Liz, and have had one for years." Nancy seemed upset again.

"The cost management system you have right now, Nancy, is just a download of the standard cost budget. Accounting then says to operations, 'This is the standard cost you budgeted for this individual part. Do not exceed it, or else!' This becomes what we accountants call 'control' but is nothing more than what operations calls 'variance analysis paralysis.' The 'control' in our Lean business now will be the Lean cost management system inside each product family, and the only variance analysis will be a variance of this month's actual costs to last month's actual costs and the year-end target cost."

"I like this, Liz!" Mark commented. "So now, instead of the accounting police, we have what I would call a product family 'Neighborhood Watch'!"

"Yes, Mark, and the new Lean cost management system also focuses not on numbers and transactions but on processes that create flow—another principle of Lean. So we have moved furniture, yes, to create product families in which specific cross-functional groups of our people manage certain groups of products and certain groups of customers. These families, which create value, and the design of operations, which creates flow, now become the prerequisites for our Lean cost management system. We no longer need an elaborate IT system and cost allocations to track costs. Our individual families will track costs—and they are all direct costs as the family is directly responsible and held accountable for the costs in their family. They can do this simply in Excel, and then show results in graphs or visuals for revealing trends."

"So what is your point here, Liz? What is the bottom line?" Jim asked.

"Accounting does not create this Lean cost management system, Jim. OPS people create the LCMS. After all, it is for OPS people! So my point is that changes in operations come first, then we link all of this to cost within the operational families to create the LCMS. The principles of Lean and the design or layout of OPS to create flow come first, then the cost management system. The LCMS is dependent on the transition from batch manufacturing to Lean one-piece flow manufacturing. This is very different from any traditional cost system! How many times has Tom asked me if I am going to sit up in my ivory tower and move numbers around in spreadsheets or come down into the plant and look at the way 'the work' works? The bottom line, Jim, is that if you are still in batch mode, you cannot implement an LCMS. An LCMS is not a subset of the traditional financial G/L! It is not even what accounting would refer to as a traditional cost system. It is not a CD-ROM or some software program that one just installs on a laptop. It is not for accounting. It is for operations! It is a reflection of the changes in operations, which in turn affect our costs."

"I would agree with Tom," said Mark. "It seems that financial statements today portray a business from at least 50 years ago. It is no wonder people in OPS get disgusted."

"Whether this business is a success or not goes back to your dog training, Mark," I said. "You told me, when I tried to train the dog, to use positive reinforcement and to not say 'no' too often. You told me that, if the dog could not find the item that was hidden, to guide him to it. You said that you wanted the dog to be successful. Does this not apply to people in business as well?"

"So what does this informal Lean P&L, or 'planning tool,' look like, Liz?" Jim asked.

"First of all, the product family is responsible for both revenues and costs, so the Lean P&L begins with revenues. This is different from the traditional cost centers we had in our first Lean attempt. Then I asked the A-team to determine the categories of cost as they see them. They started with material costs—what they actually buy and use in the month. Then they decided to break labor costs into factory labor and salary labor, with benefits on yet a third line. They will also track the cost of supplies as well as services, where supplies are items like oil, work gloves, etc., and services are things like the cost of outside vendors. A noncash cost they will track is the depreciation of machinery. Then, too, they will track the costs of utilities as well as scrap. Thus, the Lean P&L for a product family will look like this schematic here (Table 11.1).

"You should note that this Lean P&L contains essentially only two categories of actionable costs: material costs and processing costs. Rather than get wrapped up in segregating increases or decreases in the material and labor content of inventory here in the Lean P&L at this early stage when we have so much inventory, we decided to keep increases or decreases in inventory in total as a line item below total cost, then also track inventory turns as a Lean metric until we get the level of inventory down to something manageable. To avoid any allocations at all,

TABLE 11.1

The Lean P&L for a Product Family

Lean P&L, A-Team Product Family (accounts in thousands)	January Last Year	January This Year	February This Year	Year-End Target
Revenues	$300,000	$305,000	$310,000	$350,000
Actionable costs of production				
Material purchases	$160,000	$140,000	$125,000	$110,000
Factory labor wages	60,000	46,500	46,500	46,500
Factory salaries	16,000	14,000	14,000	14,000
Benefits of wages and salaries	36,000	32,000	32,000	32,000
Supplies/maintenance	6,000	5,000	4,000	3,500
Services/outside vendors	8,000	6,000	5,000	4,000
Depreciation (machine)	7,000	5,000	4,900	4,500
Utilities	25,000	23,000	22,000	15,000
Scrap	7,400	4,000	3,000	0
Total actionable costs	$325,400	$275,500	$256,400	$229,500
Decrease/(increase) in inventory	(18,000)	(12,000)	(2,000)	$20,000
Cost of sales	$307,400	$263,500	$254,400	$249,500
Gross profit (loss)	($7,400)	$41,500	$55,600	100,500
Gross profit (loss) %	−2.47%	14%	18%	29%

such as an allocation to charge the family with floor space utilized, we also decided to keep what accounting would call an 'occupancy cost' or 'building expense' off of the Lean P&L and track the family's square footage utilized as a Lean metric. Square footage is really what we want the product family to reduce. So what we have here is a 'linking' of actual cost data with physical operations. No standard cost, and no cost allocations! The 'a-word' from here on will be a 'no-no.' Operations people do not want to see any allocation of costs in their Lean P&L!"

"So who is responsible for all of the other costs outside of the product families: the R&D engineers, our Learning Kaizen facilitators, our commodity team, our executive team, our corporate finance team, and the shipping team and sales team, as well as building maintenance costs?"

"Management is responsible for those costs, and they are only about 5 to 10 percent of the total cost of the business. These are very small groups of people who serve the entire business."

"And where is the product family going to get the costs in the Lean P&L?" Nancy whined.

"The family can find these costs themselves—in HR records, in the fixed asset system, and in accounts payable or from invoices they have in their possession. These are actual costs, not standard costs, and there is a functional representative from HR and accounting in each family. Also, is it not operations who usually give these numbers to accounting in the first place?"

"So another major difference in the structure of our new organization is that we are now essentially a decentralized company rather than a centralized entity," Helen commented.

"That is true, and I believe you will see a decrease in the amount of bureaucracy in our decentralized structure. And that, too, is a good thing! But the basic premise here is that the design of Lean in operations and the implementation of the Lean principles of value and flow come first, then the Lean cost management system. And this truly is a system as we have linked cost to operations inside a family. So I believe my conference friend was correct: 'Structure drives behavior which creates a culture,' and the LCMS ensures that this new behavior sticks as we go forward with the principle of pull and our Lean improvements. No one in operations will be able to 'play the game' of absorption accounting from this point on."

As I was about to get in the car to go pick up my mother, Mary, our intern from the university across town, approached me.

"Hi, Liz. Would you have a moment? I would like to speak with you."

"Well, I am in a bit of a hurry to get to my next stop, Mary. What is it?"

"I have been listening to all that is going on here now, and I have some questions that I would like to ask. I know that our CFO does not like us to be away from our desks during work hours, and I do not want to upset her or make her angry. I need this job to help with the costs of my final year at the university."

"Sure, I understand. Would you be available before work begins tomorrow morning? I could meet you, let's say, at 7:30 in the company cafeteria for a coffee. Nancy could not object to that, could she?"

"No, meeting for coffee tomorrow morning before work would be great, Liz. Thanks! I look forward to talking with you."

12

Mary, Our Intern, Speaks Up

Accounting higher education has been warned repeatedly that the current model is broken, significant change is necessary, and a new education model, consistent with global market expectations for their student products, must be developed. This new model is required if accounting education wants to continue to achieve the business community's respect and to be perceived as adding value to the profession. ... Accounting education is at its best when it's at the leading edge of what is practiced in industry. [We need to] change the accounting education paradigm.[1]

I met Mary in the company cafeteria early the next morning. She seemed nervous. Perhaps our goal of changing behavior is not coming to fruition in the accounting department. Mary came to us highly recommended by the university for her excellent GPA and interest in manufacturing. I must admit that I am curious why a student would be interested in manufacturing as opposed to the service jobs in auditing in the big accounting firms. Academia seems to push those service jobs, as if to imply that manufacturing is not only dead in the United States but also a dirty, nonglamorous job. Even our shop workers, who have made a very good living in manufacturing, are encouraging their sons and daughters to go to college and reposition into service jobs.

"Hi, Mary! Good to see you! I am glad you could make it a little early this morning. So what is on your mind?"

"Well, I have been taking my lunch down to the A-team break room every day and listening to what the members of that team are saying. I find this new business structure, as well as the new Lean P&L, most interesting, Liz. We do not read anything like this in our accounting textbooks or hear about Lean in any of our other classes at the university. I would like to ask if I might be assigned to the next product family."

I decided to play devil's advocate here. "May I ask why, Mary? Why would any student be interested in manufacturing today when you could work in a nice office, with well-dressed, stressed-for-success auditors, just waiting for that CFO position in insurance or banking to open up and provide you with a huge posh corner office in Manhattan? I am sure that office looks very different from my office here."

"May I ask why you chose manufacturing, Liz?"

"I suppose my choice started as a child when my dad would take us on weeklong educational vacations each year. I especially remember him taking my sisters and me to Hershey, Pennsylvania, to see how chocolate is made. That was fascinating! Then, too, my first degree is in mathematics, so I have always loved solving problems. But the final fork in the road to a professional career in manufacturing came when I went back to college to earn my master's in Accountancy. I was all set to go the CPA route, into a service career, when I met a young man who worked for Toyota in Japan in one of my evening classes. He really spiked my attention in Lean in general, and then invited me to attend a talk given by his boss one evening at a neighboring university. It was like attending a religious revival meeting, Mary. Really! I was converted that night, and the next day I went back to the university to speak with the professor who was my adviser, to ask if he could introduce me to anyone in the company that I heard that night had hired this Lean consultant. I was hired by the CFO of that business, and I have been in Lean ever since."

"I think I want to do the same thing, Liz. If my experience in one of the product families here excites me as much as I believe it will, I would want to apply for a permanent job here in this company after I graduate."

"Today businesses hire the best and the brightest, with the highest GPA, and you are all of that, Mary. But then businesses take these very bright graduates and tell them to 'go get' that 'number'—as if they were throwing a bone for a dog to fetch. The 'book smart' student will do whatever to maneuver the data and reach the desired numerical result. This is what academia has taught them! Personally, I would want the student who has interned in a real business today—someone who has learned by experience. Why? Because what we teach in the classroom today in no way resembles what is happening in world-class business today."

"That is exactly what I mean, Liz! I want to learn Lean and Lean accounting by participating in the reinvention of this traditional company. That would truly be a learning experience, and something distinctive to put on my résumé. I have heard the jokes (and maybe they are not jokes) where the CEO asks the CFO what operating income looks like this month, and the CFO replies, 'What would you like it to be?' I do not want that kind of career, Liz. I do not want to learn how to manipulate data to get costs or profits to be what the CFO or any executive thinks they should be."

Again playing devil's advocate, I asked, "Can you give me an example of what you are saying occurs in your classes, Mary?"

"Sure. All through my accounting principles courses we have been taught that WIP is a very important inventory account, for one example. It seems to me that every transaction in Accounting Principles I includes a WIP account! But the Lean philosophy of business teaches that inventory, including WIP, should be very low, and we should reduce transactions which are just another form of waste in addition to the waste of inventory or WIP itself. Even the Lean principle of flow tells us to eliminate not just batches but WIP!"

"You do have a point there, Mary. Perhaps we should teach the history of accountancy in our colleges."

"What do you mean by that, Liz?"

"Think of the boot maker who owned a shop back in medieval times. He did not make hundreds of boots in all sizes and colors, and then place them for sale in a store. No, a customer would come to him for a new pair of boots. The boot maker would trace around the customer's foot, and take a small deposit which would enable him to buy the material to make the boots. When the boots were completed, the customer would come back, pay the boot maker, and go home with his new boots! His business was truly a Lean business. And I doubt that he utilized absorption accounting in his books or financial statements, nor do I imagine that he spent too much time worrying about EPS (earnings per share)."

"That is a good example, Liz. Then, too, traditional accounting, as it is taught at the university, also teaches us to rely heavily on costs allocated to individual products based on labor hours or labor dollars. But I can see in this company that material is about 70 percent of the total cost of products, which means that labor and overhead could be at most 30 percent of total cost. By relating everything to labor costs, one is definitely distorting the cost of a product—not to mention making this methodology for product cost accumulation extremely complex. But Lean says to keep it simple! And just look at the chapter in our accounting principles text on depreciation. They teach us so many methods to calculate depreciation, and then teach that we want to choose the method that will maximize cash flow rather than reflect the true economic life of the machine. I know that Lean puts a lot of emphasis on cash flow, but …"

"That, Mary, is yet another example of what I call 'playing the game.' I know they teach this in the MBA programs, but I had no idea that they teach this game-playing as early as Principles I. As most professors have never worked in industry, perhaps they do not see this game. Henry Ford told us back in the early 1900s that a machine is worth only that which it can contribute to the manufacture of the product. I tend to see his point and yours, Mary. I do not believe I could work in a traditional non-Lean accounting position today. It makes my head numb just thinking about it! What is taught at our universities today truly is not relevant in business today."

"That is so very true, Liz," Mary remarked.

"Having worked as an adjunct at a university, I can vouch for much of what you are experiencing, Mary. Perhaps the solution is for the university to Lean itself! Please allow me to digress. I was a guest lecturer on Lean in an MBA class not too many years ago. One night, after a short lecture, I had a rather complex case for the students. The professor wanted to perform an experiment with this case. So he divided the students into three groups. One group was composed of all students with financial undergrad degrees. A second group was composed of all students with engineering undergrad degrees, and the third group was made up of a mix of students with both financial and engineering undergrad degrees. The professor and I walked around, listening to how the three groups were attempting to solve the case. The group of financial students was saying, 'This is too hard! What has this got to do with business? There are no numbers in here!' And in the group of engineering students, each student had his or her own solution and was arguing over whose was best. But then the cross-functional group had the near-perfect solution to the case—and in record time. They were having fun! The cross-functional 'family' approach to problem solving seemed to work the best—just as it does in industry."

"I see your point, Liz. We need to have classrooms that simulate the real world."

"That is a good way of putting it, Mary. The university is teaching students to optimize the parts of a business—the financial piece, or the engineering piece, or the management piece. But we want to create a system where everyone is linked and working together, and then optimize that system. We do not see this type of thinking encouraged in our textbooks or in the classroom. I believe my classes are the only classes at the university that use the classroom with large work tables. I place all of my students in teams at those tables from the very first day of class."

"I wish my professors would allow us to work in teams, Liz. We would learn from each other."

"And one other thing I have noticed in academia, Mary, and perhaps you have too, is that there is no training or teaching in *processes,* for example, the process of problem solving. My students want a recipe for each type of problem. That is to say, they want to be given a model solution A for problem-type A, a model solution B for problem-type B, and so on. If I give them problem-type C, they do not know what to do! There is no predetermined recipe for solving problems in business. But the university does not seem to teach anything such as the Six Sigma DMAIC process (define, measure, analyze, improve, and control) or PDCA (plan, do, check, act) process—both of which are utilized extensively in business today for problem solving. I believe that students have never even been taught how to learn."

"Did you 'learn' how to learn in business then, Liz?" Mary asked.

"Perhaps I did, Mary. I distinctly remember my initial job in a large manufacturing entity. My first week in the business, I was called into the office of one of the VPs who managed an entire product line and OPS group in the plant. He gave me my very first work assignment my first week on the job. This first assignment in an industry I knew nothing about was to write a business plan to take 8 percent out of the cost of one specific, mature, multimillion-dollar product, and then put together a team to make that plan happen. He did not place me in a training class! That project got me into the plant to observe operations and talk with OPS people about what they believed the problems were. That project also sent me out to suppliers to learn their businesses. And then the execution of the plan introduced me to VE (value engineering) and scientific methods of problem solving which were in addition to what I had learned in my mathematics classes at the university. That was the best encounter on how to learn that I have ever had in my entire life, Mary. It was truly learn-by-doing as well as trial-and-error."

"Well, I think you have definitely reinforced my choice to be a part of the business solution here in this company, Liz."

"And I believe, after speaking with you, that this would be a great opportunity for you, Mary. And perhaps this would be the start of something great, as well as exciting, for you in the future. Did you know that a CPA or CMA with a Six Sigma Black Belt can earn well in excess of $100,000 a year—and at the same time make noble contributions and add incalculable value to a company, and have fun doing it? Perhaps we could pitch an advertisement at the university to inform students that manufacturing is sexy!" Mary smiled.

"They do not tell us this at the university either, Liz. They just push auditing careers. Thanks again for this conversation. I had better get back to work in my glass tower!" and we both laughed. Mary will make a great addition to the next product family and a great addition to the corporation after she graduates. We have found a "Lean thinker" in Mary!

I cannot imagine how students such as Mary do not get totally frustrated with the curriculum at our universities today. Academia today is still teaching Sloan's philosophy of business in which the goal of the business is to make money, period. We do not teach the Lean philosophy of business in which the goal is to provide a service for the customer and contain costs via waste reduction. Henry Ford told us that "... our recipe for 'hard times' is to lower prices and increase wages."[2] Do we teach this today? I think not. Faculty are required to teach traditional 1930s accounting as no one seems willing to update textbooks today so that the topics in the texts align with what is going on in business today. But we who teach have an obligation to ask our students, after teaching the basics from 1930, "What is wrong with this picture?" One small university requires that all students who want admission to the College of Business be required to enroll in, and successfully pass, Accounting Principles I and II. I believe that these courses, taught in a fashion that asks students to research and challenge what they are being taught today, would satisfy the most sought-after course requirements in "critical thinking" and "problem solving" which our students today need so desperately. A few accounting professors have been talking about changing the current education model to teach problem solving at our universities, yet no one seems to hear these professional people who have been inside industry.

After meeting with Mary, I ran over to Mark's office. I hoped to catch him before he got totally engrossed in his day.

"Hi, Mark. We have not talked in a long time! How are things going with the A-team and the design of the flow for the family's future value stream?"

"It is going great, Liz. The family has so many ideas for improvements in its business, and they are so excited with the way that management has empowered them to make a difference. This also is giving top management exposure to what could be in our business, and helps them in developing a plan to prioritize what comes next."

"That is good to hear, Mark. We really have done things differently than before. This time we are teaching the Lean principles—not just Lean tools—and we truly are the embodiment of a learning organization. People are learning about the Lean principles, and then they are given time to actually implement these principles into their piece of the business. They are no longer reading books, nor attending seminars or some such training off-site. They are learning and immediately putting into action what they learn. The 5S Learning Kaizens got our people to take ownership and then, linking knowledge they received the first two days with action to implement in the final two days, allowed people to voice both positive and negative opinions, which created trust between our associates and management. I love seeing management listening in awe to what the OPS people are telling them. Just like you have taught me with your dog training, Mark—positive reinforcement, practice, and party! OK, maybe *celebrate* would be professionally more fitting here. And this is a continuous process, not a once-a-year event."

"That is true, Liz. I meet with them every week, and we update the value stream map each time we make improvements. But we will need the Lean cost management system (LCMS) and some Lean metrics to validate the cost savings we are seeing."

"I am so glad to hear you say that, Mark. One prestigious university that has a Lean program is teaching the use of ABC (activity-based) costing as a method for validating the cost savings of Lean initiatives. Why this is not true is a topic for another day. However, just think for a moment about what you know of ABC costing in general. It certainly does not focus on simplicity. It is a complex, huge software program that requires an army of people to develop and maintain the system. It is yet another system 'by accountants, for accountants.' Does it address responsibility and thus empowerment? I think not. Also, it does not involve all associates of the business, and it certainly does not involve those people who do the work, thus it will not help us with continuous improvement or operational control. It is never going to change the way people think! And it can be manipulated by the financial people who put it together. But foremost, ABC is an allocation system, and we do not allocate costs in Lean. Why would we utilize such a cost system to validate our Lean improvements? This adjunct's presentations made me want to tackle him and throw him out of the window! Fortunately for him, the windows did not open. The LCMS will give us the baseline we need to trend all the good things that this one family has been engaged in to date. Once the A-team is comfortable with everything, then we can move on to the next family. We are all set with the members of that next family, are we not, Mark-san? Oh! That reminds me. I had a very interesting conversation early this morning with Mary, our accounting intern. She would like to be assigned to the next family, and I believe she would be a great contributor to that family. What do you say?"

"Sure! Why not? She may bring some fresh new ideas."

"Thank you, Mark. Are we ready for the Lean metrics meeting this afternoon?"

"We are! We have no metrics! We will have to convince them as to why we need this—just as you had to convince them that we need a Lean cost management system. Actually, it seems to me that a metrics scorecard will be a part of this Lean cost management system. Will it not?"

"You are correct, Mark. The Lean P&L, the Lean metrics, and some other pieces which we have yet to discuss, constitute the Lean cost management system. See you at the off-site!"

REFERENCES

1. Russell, Keith, and Carl Smith. 2003. "Accounting Education's Role in Corporate Malfeasance: It's Time for a New Curriculum!" *Strategic Finance*. (December), 46–50.
2. Ford, Henry, and Samuel Crowther. 2003. *Today and Tomorrow*. New York: Productivity Press, pp. 219–220.

13

Lack of Metrics Is a Recipe for Failure

Most business measures today are worthless. Why? It is because, just like our traditional standard cost system, these current measures are created "by accounting, for accounting" but should be created "by OPS, for OPS." We measure whatever is easy to measure and not what we want to improve—what will validate our Lean progress and align with our Lean business strategy. Yes, at the start of our Lean transformation, we do not have the culture in place for new Lean metrics. But then, too, most of what we currently measure is not utilized in managing the business, and what we currently measure is most likely not something OPS people can influence or improve. What we do have is merely an exercise in micromanagement as opposed to managing the whole system—and with a plan to meet the targets.

The Steering Team, including Jim, Nancy, Helen, Tom, Mark, and me, and also Greg, our Lean leader, came together after lunch at a local hotel conference center. I am going to stay in the background this afternoon, and let Mark and Greg lead this discussion.

Mark began the afternoon discussions. "We want to motivate people, change behavior, reduce waste, and improve processes to better serve our customer, ultimately increasing both revenues and profits and growing the business through acquisitions—this is what we desire to achieve with our Lean transformation. This is our strategy. Would everyone agree?"

"Yes, definitely," responded Jim.

"So we should measure what we want to achieve, and not what is easy to measure. Our new metrics should help us gauge our strategy as well as our overall company performance through this Lean journey, and evaluate how we are doing in implementing our Lean strategy of business. Would everyone agree?" Mark inquired of everyone.

"Yes, of course," responded Tom.

"If we only measure financial results, we are apt to get dysfunctional behavior. We have seen examples of this just recently. We measure and place great emphasis on PPV (purchase price variance), and so we have buyers who purchase cheap, low-quality boxes in order to get the financial result that will give them a bonus or a favorable report card. Does everyone see this?"

"Yes, I do," said Nancy.

"Then why, given all of the above, do I see only 'cost per case' as a metric reported in our monthly financial package in this business?" Mark asked rhetorically. "And why, when someone reported the monthly financial package to my OPS group, did someone in my OPS group have to ask what was in the calculation of 'cost per case'? If one has to ask what makes up the metric, it is probably too complex. Would you agree?"

"This metric of 'cost per case,'" Nancy replied, "is to help us control costs."

"But," Greg said, "Mark just said that the purpose of our metrics should be to create the right behavior in our people, improve processes that create value for the customer, and align with our corporate strategy for growth. Yes, a metric might create a control mechanism, but if no one knows what is in the calculation of the metric, how can this help? Besides, do we not want metrics that assess how well we are doing in our Lean progress?"

"And that would then imply," Tom added, "that our metrics should be not only financial in nature but rather mostly nonfinancial in nature. Financial measures are *lagging* and not *leading* measures. Financial measures are *outcome* metrics. We calculate these financial measures from our financial statements, which contain figures that are four to six weeks old. Thus these financial measures are no indication of the future success of the business. Running a business based on financial measures would be like driving a car while looking in the rearview mirror! Plus, financial measures are short term and reward the wrong behavior. Most importantly of all, they are under the control of the accounting department, and OPS people in most cases cannot act upon them nor change them without playing some type of 'game.'"

"I don't understand," Nancy commented.

"We will take much of the input data for our Lean metrics from our value stream maps, not from your general ledger, Nancy," Tom said. "These are the items—the processes—we want to improve. Some of these inputs into new Lean metrics might be number of employees (which is definitely not on the balance sheet), throughput, first pass yield, equipment uptime, and cycle time—all not in the general ledger, Nancy."

"And Lean tells us to 'keep it simple,'" I added. "So I think we want metrics that have just one numerator and one denominator. That would rule out a metric such as ROI, return on investment. If we have a metric such as ROI that has more than one factor in the numerator or denominator, and the metric is not what we think it should be, then how do we know which variable to change to improve the metric? Which button do we push to make the measure better? We don't know!"

"You have a good point, Liz," Mark said. "Let's take productivity as another example. I believe we would all agree that this is one metric that we need in order to track our Lean progress. And this would be one metric we could take from Toyota, which is a productivity monster! Productivity, simply stated, is sales per employee. So that implies that we have one factor in the numerator, revenue, and one factor in the denominator, number of employees. Simple arithmetic tells me that, to increase this fraction or ratio, I must either increase revenue or decrease number of employees."

"Yes, and I do not think we want to decrease the number of employees," Helen inserted immediately. "We have pledged that Lean will not equate to layoffs."

"That is right, Helen," said Greg. "So the only alternative is to increase revenues. But *how* we do that is important, too, Helen. Do we increase revenues by increasing prices, or do we increase revenue by growing the business?"

"I see your point," Jim said.

"And let's take two other examples of metrics," I responded. "Let's look at two very popular metrics: operating income and inventory turns. The first one, operating income, is purely financial in nature and the second one, inventory turns, is more operational in nature. The first one is short term, and the latter is long term. But more important is the fact that the two are mutually exclusive! If we have both operating income and inventory turns as metrics, we are creating dysfunctional behavior—at least for now."

"I am not sure I get it, Liz," Jim replied.

"To have good inventory turns, Jim, at least for right now, we must use up all of this stash of raw material and WIP we presently have. However, to have good operating income—at least for now—we would continue to buy more and more raw materials, make more than we know we can sell, and defer costs of products we do not sell to the balance sheet as inventory. And that would just lower our inventory turns. So, you see, you cannot have it both ways! And, as you and I discussed earlier, using up current stashes of material implies we may show a loss in operating income as long as we are drastically working to lower inventory. So if you ask your product family managers to increase both operating income and inventory turns in the early stages of Lean, you are giving them an impossible task, and you will get more dysfunctional behavior, and more game-playing."

"I see, Liz. I guess I forgot about that discussion we had. So we would want to go with the long-term metric of inventory turns rather than the short-term financial metric of operating income, correct?"

"Yes, that is correct, Jim."

"OK, so we have productivity and inventory turns as two of our new Lean metrics. But the factors that make up those measures are financial in nature, so both of these metrics would be considered to be financial measures. Would everyone agree?" Mark wanted consensus.

"I do not believe that I would necessarily classify productivity as a financial measure," I argued. "It is my understanding that productivity is defined as finished goods per employee per day. I would then see this as a *process* metric rather than a financial metric because OPS people can act upon and alter both factors of the metric: finished goods and number of employees. And I believe the same would be true for inventory turns. OPS people can impact both cost of sales and inventory, so this metric is also more of a process metric than a financial metric."

"That is a good point, Liz. So what other metrics should we be concerned with in order to track our Lean progress?" Mark inquired.

"I once heard a very successful Lean CEO say that, if he had to choose just two metrics to monitor his company's progress in Lean, those two metrics would be inventory turns and customer service because if you are reducing inventory and focusing on what the customer wants, you cannot help but take market share and grow your business. Customer service is not a financial measure, is it? I believe I would consider it to be a *customer* metric," said Greg.

"I would agree that the customer needs to be in the Lean metrics. But how would we measure customer service?" asked Helen.

"Good customer service implies giving the customer what he wants (that is VE, value engineering, and QFD, quality function deployment), at the price he wants (that is, target costing), and when he wants it, which implies well-executed delivery. So we might measure customer service as shipments shipped correctly the first time divided by total number of shipments. Call that on-time delivery. Or the measure might be in terms of late shipments, in dollars and in number of parts, or a customer return percentage. Or we may even want to measure customer service in terms of lead time quoted to the customer. Call that order-to-delivery. A percent reduction in customer complaints might be yet another measure of customer service; however, I think that would be a lagging measure of outcomes. I think we want more leading measures that drive performance." Mark spoke from experience.

"I may be off the track here, but how do we decide what categories our metrics will fall into? Liz just essentially mentioned QCD—quality, cost, and delivery—as the categories to measure. But I always think about Toyota's True North pillars in relation to metrics. These pillars of people and processes need to be addressed and monitored in our new Lean business, do they not?" Jim inquired.

"What is Toyota's True North?" Helen asked.

"True North," Greg responded, "is Toyota's vision, values, and metrics all succinctly wrapped in one package and tied with a bow! True North includes one, developing the *learning* skills of people, including problem-solving skills; two, improving *quality*; three, reducing the *cycle time* of production for increased speed to market (that is, a *process* metric); and four, increasing profitability through increases in *productivity*—and not through raising prices. Note that they do not really have any financial metric. These four key components of Toyota's True North metrics drive its Lean improvement initiatives while, at the same time, linking these nonfinancial metrics to the more vital financial figures in the company's business strategy. They do this by ensuring that each metric has a cause-and-effect relationship. It is a model to be respected, but not necessarily copied."

"Bottom line: we want our measures to be simple—Lean tells us to 'keep it simple'—and easy for everyone in the company to understand, and we want the metrics to focus on processes and the flow of material and information rather than focusing only on numbers and financial results. Just like with the LCMS, we want the metrics to be OPS based (not accounting based) and align with our company values and strategy, and to motivate people to do the right thing," I added.

"I agree. And our choice of inventory turns is a measure of material flow, while our choice of a customer service measure (such as order-to-delivery) is a measure of information flow."

"OK, so we have decided and all agreed upon three metrics thus far: productivity (a process metric) and inventory turns (also a process metric), and something that we would call a customer measure. Customer satisfaction implies good customer service, which is really on-time delivery. This measure is simply the number of shipments shipped right the first time divided by the total number of shipments. We do not want customers complaining that they received the wrong items in their order. Or customer satisfaction might insinuate how well we do with quoted lead time, which is essentially order-to-delivery. We are good at this order-to-delivery. However, we do well at this because we keep so many raw materials in stock. As we get Lean and reduce our inventory of raw material, will we still be able to deliver at a faster rate than our competitors? I believe we want to continue to monitor this to ensure we stay above our customers' expectations. What else? Are there other categories of metrics that we need to ensure that we stay on track with our Lean progress?" Was Mark really seeking additional categories of metrics? What is Mark trying to get us to come to realize? What is he trying to teach us?

"Quality is one of our company values, and so it should be a metric also, and it is 'True North.' It is very important! And I would categorize this as a process metric," said Tom.

"OK, but how would we measure quality?" asked Helen.

"Quality could be measured as defects per part or first pass yield. Or we might define quality as a percent reduction in rework," Greg explained.

"I believe first pass yield makes more sense for our business in measuring quality," added Tom.

"What about measuring cost?" Nancy whined.

"We are measuring cost in the metric inventory turns, which is defined as COS (cost of sales) divided by average inventory," replied Mark in a subdued fashion. "Liz merely commented that we should categorize this as a process metric as OPS people can impact it."

"OK, then we said we would track the amount of space each family works in as opposed to allocating an occupancy cost to each Lean P&L," I reminded everyone.

"And what about tracking capacity as a metric? Our vision is to grow the business and our market share through acquisitions and new products. So we must keep an eye on new products as a percent of sales as well as the amount of capacity we open up as we work faster and smarter. There will be a point in time where we want to take some of the cash we accumulate and use it to buy another business to drop into available capacity. We will not want that freed-up capacity to sit idle for too long," Jim advised.

"And might we want cycle time as a metric? This is also True North, as well as a process metric. This is quite different from lead time as a customer service metric. If we have plentiful inventory, our lead time is certain to be excellent. But cycle time is more like process time. It is the amount of time it takes a part to pop out of any process box on the value stream map, but it does not include the wasted time a piece of WIP spends sitting between process boxes. So perhaps a better measure would be *throughput* time—the rate at which sold products proceed through the entire manufacturing process. This would include both processing time and wait (or waste) time." Mark surely knew his Lean business and added value to this dialogue.

"I believe you are correct. Throughput would be the better process metric," Greg added.

"Do we even have a financial metric here?" Nancy truly is one of many people in a business who tend to gauge the success of a business strictly from looking at financial measures. However, do we not desire to get to the point (as evidenced in True North) where executives will realize that it is the nonfinancial metrics that truly drive the business and, via cause and effect, are the leading metrics that bring about success in the lagging financial metrics? Perhaps this is what Mark is trying to teach us.

"Sure we do! We have inventory turns, which is COS divided by average inventory, both financial figures. And we have sales, another financial figure. Then, too, productivity contains finished goods in its calculation—another financial figure. While we decided that COS and inventory and finished goods are items that OPS can act upon and change, they are dollar figures. Perhaps, based on what Jim had to say, we should add a financial metric that would measure the percent of sales coming from new products to gauge how well our R&D folks are doing with new products." I tried to respond tactfully, and at the same time bestow a little fulfillment and contentment upon Nancy.

"Growing workforce skills, better known as training, is another True North metric. Do we not want to include employee training or learning as one of our Lean metrics?" asked Greg.

"I think we have an excellent all-inclusive Learning Kaizen program in place for employee training. Perhaps our learning metric might be hours spent in Learning Kaizens. We told our associates earlier that they would be required to take part in two weeks of this training, which links learning with action each year, and that equates to eighty hours per employee per year, or we might measure number of employee suggestions implemented as a result of time spent in a Learning Kaizen. Or we might even want to track how many volunteer hours an associate spends in the Learning Kaizens. Our Learning Kaizens are teaching our associates that predetermined solutions to problems are not acceptable, but rather we learn by doing and making mistakes. So I do believe this is an excellent metric for gauging how well our people are learning." I strongly believed this was a better metric than hours in book clubs.

"OK, but what about safety as a Lean metric? Safety is very important. It impacts, or could impact, all of quality, cost, and delivery, could it not?" asked Greg.

"I think you have a point, Greg. We should add safety to our Lean metrics. And I believe that this would fall into the category of a learning or a process measure," Mark concluded.

"Do we have scrap as a metric? We have never had a system in the past to track scrap."

"I believe we need scrap as a metric, and also we need to track employee turnover. All of those persons who could not or would not participate in change left the organization at the start of our Lean transition. Do we not want to track how many others might leave? This would alert us to the fact that our attention to our employees' growth and learning is not working," I inserted.

"So I believe Liz has been busy placing all that we have discussed and decided upon in a metrics scorecard, and here are our new Lean metrics (Table 13.1). Notice that we have discussed three models for the metrics scorecard. One is Liz's model, which she calls QCD: quality, cost, and delivery. Toyota's four pillars model has as its categories of measure: quality, productivity, cycle time, and learning. This is the same as Liz's QCD model with employee learning added—let's call it QCDE. Then we have a popular scorecard developed by a professor at a leading college which has as its categories of metrics: process, financial, customer, and employee learning. I believe all of these models are very similar, and any differences in categories of measures are of little significance." Mark was beginning to make his point, and I believe I saw it coming an hour ago. "The big question is this: 'Do these metrics we have chosen tie or link to our Lean strategy and translate into measures of our Lean operational success or failure, and do we have cause-and-effect links among our metrics?'"

TABLE 13.1

The Lean Metrics Scorecard for a Product Family

Lean Metrics Scorecard, A-Team Product Family	Metric Category	Current Period/ VSM	Period 1 or Future VSM 1	Period 2 or Future VSM 2	Year-End Target
Net sales (in thousands)	*Financial*	$310,000	$312,500	$315,000	$350,000
% of sales from new products	Financial	0%	1%	2%	10%
Quoted lead time in days (order to delivery)	Customer	14	12	10	7
Customer service (on-time delivery)	Customer	85%	90%	93%	100%
Inventory turns (COS/average inventory)	Process	2	5	8	20
Productivity (finished goods/ employee/day)	Process	20	26	30	40
Throughput (in days)	Process	56	50	40	1
Quality (first pass yield)	Process	60%	70%	75%	100%
Occupancy (in square feet)	Process	1200	1000	900	600
Available capacity	Process	2%	4%	12%	10%
Scrap (as % of total production)	Process	14%	12%	11%	0%
Safety (lost days due to accidents)	Learning	2	1	1	0
% Employee suggestions implemented	Learning	25%	30%	40%	100%
Percent employee turnover	Learning	20%	2%	1%	0%
Hours/employee in Learning Kaizens	Learning	40	0	40	80

Note: VSM = value stream map.

"I see your point, Mark. Everyone, if you will read the metrics scorecard from the bottom up, I believe you will see what Mark is trying to teach us. If our employee learning measures are improving and closing in on the targets, then we should see our process measures improving. If our process measures are improving, then that should lead to improving customer measures. Improvement in customer service will lead to new as well as repeat customers, which should lead to better sales, which is the financial measure. It all starts with an emphasis on gauging how well we do as a learning organization, and then links to other metrics, which finally result in improved sales—the financial measure. This is what Mark means when he says there should be a cause-and-effect emphasis in our Lean metrics." This is an important point, which Mark had been trying to make from the very beginning.

Then Mark continued, "And another thought to consider is this: Do we at this time have a culture that places a value on continuous improvement and problem solving? In other words, will our associates act on these metrics to improve them or will they be similar to our traditional financial statements where folks just file these scorecards in the round file? I believe our metrics are yet another chicken-and-egg dilemma, as Liz calls it. We will not be successful at tracking these measures of Lean progress unless we first have the principles of Lean and the LCMS in place in our business—which implies having the Lean culture."

"I believe we are ready for this," Jim added. "My only concern, Liz and Mark, is that people may project negative views on nonfinancial metrics if the financial metrics are not what they believe they should be. We are talking strategic controls now—not just financial controls. I myself am very pleased with these new Lean metrics."

"Jim, if we truly have linked all the metrics together, including ultimately linking nonfinancial to financial metrics, then I believe we are good to go with what we have decided upon today. I thought about weighting our various metrics to give less emphasis to financial metrics. Although, if you look at what Liz has been putting together in our metrics scorecard as we have been talking, we have only two of fifteen, or 13 percent, of our measures in the financial category. Then note also that we have four learning metrics, which represent about 25 percent of our measures. Our process metrics represent about 50 percent of the total, and our two customer metrics represent about 13 percent of the total. Learning and process measures are most important in leading us to better and improved financial measures, and 75 percent is the weight given to these two categories of metrics. Are we all in agreement that this weighting of our proposed metrics fairly represents our Lean business strategy?"

"OK, then if we are all in agreement and accept this Lean metrics scorecard as well as the way we have defined the calculation of the metrics themselves, and believe that they tie to our Lean strategic goals, each product family will compile and track not only its Lean P&L but also its Lean metrics. Again I want to emphasize that both the Lean P&L and the Lean metrics will drive the management of the business, and both have been created by OPS, for OPS. We have taken these report cards out of accounting! We will have the same reporting methodology for all families. These will be updated frequently, but definitely each time the family updates its value stream map. Thus, we have a snapshot in time of our Lean progress, and we are able to see trends as well as validate our cost reductions. I know you will object, Jim, but these Lean P&Ls and Lean metric scorecards need to be on the wall as visuals for all to see. There can be no secrets in our new Lean enterprise," I added.

"I understand, Liz, and I agree." But then Helen came up out of her chair, stating that she would not tolerate employee turnover statistics being displayed on a wall inside the business. And to that, Jim replied, "From now on, if there is truly trust among us, there is neither data nor information that is secret." And on that note the meeting concluded.

As Mark and I walked out after the others in the meeting had left, we began to talk about some additional information that we probably should have brought up at the meeting in front of the other members of the Steering Team. "Perhaps we should have added one more metric to the list, Liz," Mark commented.

"What would that be, Mark?" I asked.

"Perhaps we should have something like number of hours on the shop floor as a metric for the management team to ensure that they do spend more time on the shop floor than they spend reading traditional financial statements and computer-generated reports. We should also suggest that the members of the management team receive their bonuses, not based on their ability to manipulate numbers on the traditional financial statements, but rather on how well they meet the targets in the metrics we have set up to monitor our Lean progress."

I agreed fully with Mark. "If financial metrics, which are calculated from numbers in the financial statements, are good, then many people, namely top management, assume that the business organization is doing well. But—not always! We have already pointed out that operating income can be very good while, at the same time, the business is building inventory and headed for disaster."

"You have an excellent point, Liz. And I further believe, then, that we eventually want to get folks in accounting and all the others, especially the executives who typically rely strictly on financial data, to realize that the financial measures are the result of all of us meeting the targets in the nonfinancial measures first. But that will be a tough task to accomplish, Liz. Judging the performance of the business seems to be very subjective—even with our well-defined Lean metrics. I guess this is further proof of your chicken-and-egg theory. And by that I mean that there is a right (or wrong) time to introduce Lean metrics. We sure do have to have the Lean principles and LCMS in place first so that we can pull and not have to push these new measures. They are meant to gauge our Lean progress and keep us on track with our Lean change management initiative and strategy. However, if the Lean culture is not in place (i.e., we have not changed behavior), then the execution of these metrics will surely fail."

14

A Behavioral Revolt! Bring in the Sales Guys!

When an attorney friend of mine asked over lunch what I do for a living, I did not use the term Lean as he and others tend to equate Lean with a certain diet program, then laugh. Instead I like to refer to this Lean change management endeavor as a business turnaround—something my legal friend understands from his bankruptcy work. But here, in manufacturing, one can do everything possible in this Lean turnaround and still fail because one has ignored the salespeople or the marketing people—or all of the above! Any of these people, who technically sit outside of the product families, must be included in the employee learning of Lean behavior or they will sink the ship.

Six months have passed. Mom has returned home and has been back at work at the hospital, although she remains mysteriously quiet regarding her volunteer work in our Saturday morning phone conversations. Mark and I have worked closely with both the Steering Team and the A-team. And we have a second product family up and running! Still, I cannot help but think that something will go wrong, and even anticipate the second shoe dropping any day. Will Jim back down and stop this Lean initiative again? Will Nancy undermine the Lean P&L and Lean cost management system? How would Mark or I react to either crisis?

Mark came rushing into my office one day after work. "Look at this!" he said.

"What is it, Mark?"

"This is the Lean P&L for the A-team for the past six months. Look at the visual of the factory wages. These wages have been up and down, and up and down, since the family began. When I asked someone in the family about this, I was told that they had to hire some temp labor to keep up with production, and then lay the temp help off when sales fell back down. Then they ramp up again! This is killing productivity, Liz. But when I look at the production plan, I see that sales are pretty much flatlined across the first six months. How do we explain this?"

"Well, it seems to be a problem with the production plan and/or the sales plan. That is for certain."

"I looked at the traditional income statement for the entire business, and there is no indication that anything is wrong in that financial statement."

"Of course not, Mark. The A-team Lean P&L will have much more detail for this specific problem. I guess the good news is that if you do not see the problem in the traditional income statement, then this problem must be confined to the A-team product line. And we would never see this segmented detail of information in the monthly traditional income statement. Financial gurus always place a product line with very low margins together in a P&L with a product line with very high margins. That is yet another 'game.'"

"I just had another scary thought, Liz. We really did not bring the sales guys into our Lean kickoff efforts or our Learning Kaizens. Perhaps we have a behavioral problem with the sales team?"

"I think you may be right, Mark. I just knew that things had been going too smoothly with this Lean transformation. I have been so concerned with Jim and Nancy raining on this parade at some point that I never gave any thought to the small group of sales guys who sit outside of the family but do impact operations in a very big way."

"Let's bring in the head of the sales team and one or two other influential sales guys for a Learning Kaizen. We need to get to the bottom of this, Liz, and correct any dysfunctional behavior before we have an even bigger problem. If you can spare some time to join us, I will get three of these sales guys—those who can influence others—in here for day after tomorrow. And I believe it might be a good idea to have Mr. Sato lead this particular kaizen."

The sales manager plus two other sales reps were brought in as planned. They were given the same Learning Kaizen instruction on vision, value, and flow as all company employees had received. The Lean experts say we often neglect to include the financial people. How could we have neglected to include the sales guys? After a day of training in the Lean principles of value and flow, the company mission and vision, among other topics, we had the sales guys observe production on the plant floor for a day. Then we brought in Mr. Sato.

The results of this Learning Kaizen were what I had expected. We had overlooked the sales team in forming this new Lean management system, and they were still working in batch mode. The objective, in their minds, was to sell as much product as they could to optimize the commission they earned on sales. Their primary goal was to beat the sales plan, not meet the sales plan. They were on Level 5 of our HR systems of merit. And so our A-team production squad had to ramp up and down to meet the unexpected spike or dip in sales each month.

Mr. Sato instructed all of us on a Lean concept that someone translated to be "leveling." We knew we needed to level scheduling until we succeeded in having zero changeover times. But we also needed to level selling! This meant that the sales guys needed to make an attempt to create long-term relationships with our customers so that surges and dips in production could be eliminated. The sales team can still sell a million products more than the forecast for the month. They just need to make arrangements with the customer to deliver the order in increments that will not disrupt the production plan as set forth in the operating budget.

With that problem solved, Mark and I breathed a sigh of relief. "You know what this means, Mark. From a financial perspective, I believe we need to change the way we do our budgeting for operations."

"What is wrong with the way we budget, Liz?"

"For starters, it takes too long. We start the budgeting process in summer, and are lucky if we complete it by year-end. What kind of budget is that? For half of any year, we have no budget! Then, here too, people 'play the game,' and 'the game' this time is to budget more than you expect to spend (or sell) because you know that management is going to come back and tell you to take another 5 percent out of your plan. Obviously, too, the sales guys do not take this budgeting exercise seriously. They do not give any thought to how the number of units forecast each month impacts operations' plan for labor in its budget as well as the impact they have on the production plan overall. "

"OK, I agree, Liz. So how do we fix this problem?"

"Well, first of all, there is little communication between OPS and the sales guys. Bringing the sales guys in to see operations on the shop floor is a step in the right direction. However, I believe OPS people, Mark, have some responsibility here, too. OPS people, who traditionally have worked

24/7 making product to cover overhead, have always had the philosophy of 'we make it, now you sell it!' This is all just part of the Lean learning we are experiencing. People will come along as we continue our Lean learning journey. It is good that we are discovering these things ourselves."

"OK, Liz, but again I ask, how do we fix it?"

"We fix the traditional budgeting process by locking off the standard cost system to any persons outside of accounting, and replacing the traditional standard cost budgeting process with target cost management!"

15

Target Costing for Profit Management

Target Costing is a move away from the traditional, financial, department-focused accounting system to a more process-oriented, managerial, cross-functional cost system which is, by definition, OPS-based and not general ledger-based. It involves collecting data not found in traditional financial cost accounting systems, for example, life cycle costs and costs of processes in value streams. In an ABC (activity-based) cost system, which some in academia believe is a Lean cost management system, one is mapping costs found in the general ledger to certain activities in cost centers. In target costing, however, a lot of the cost data you need will be found anywhere but the general ledger. To make target costing work for you, you must be a proactive, think-outside-the-box, new-age, world-class management accountant who is, first and foremost, a team player. Moreover, target costing is a cross-functional process for involving marketing, R&D engineers, commodity management, sales, core finance, and top management executives—all of those functions that typically reside outside of the product families—and linking all of them with operations. It is a system for profit management.

Mark came to my office after work that afternoon to discuss our operations budget. "The OPS budget all commences with the sales forecast," Mark began. "And that implies that we are not just interested in the accuracy of the prediction of number of units to be sold in any given month—although that seems to be a major problem right now. It actually begins with QFD—quality function deployment—which is essentially the voice of the customer. And so it seems to me that we need to get another group that sits outside of the families, the marketing guys, to alter their approach and replace the traditional *cost-plus* price with a *market-based* price."

"And the simplest way to do that, Mark, is to lock off the standard cost system. The marketing guys depend on the standard cost for material and labor and overhead to price all of our products. You engineers, however, do understand QFD as well as target costing, which is essentially a cost management process for reducing the total costs of new products at the planning and design stages, and their use in pricing products. Here you guys involve engineers and outside vendors from whom we purchase materials or parts. So, for starters, I believe we need to also involve the marketing guys and those persons in the product families, including the financial folks, in this QFD team, as well as the VE (value engineering) team," I commented. "These teams are not just for engineers, Mark. These teams need to also include accounting, marketing, procurement, and operations people as well as engineers."

"Sure! Can do, Liz! Then how does all of this fit into the actual budgeting process as you propose to Lean it?"

"What I would call *target budgeting* utilizes the same concept as target costing. Target costing begins with the process of determining sales prices that align with the voice of the customer, then setting highly aggressive target costs that are really *stretch goals* and inspiring employees to have a plan (or process) for how they will reach those targets that are most aggressive. In this manner, with the help of both VE and QFD, we now have built into our products both quality and functionality, at a price the customer likes, and thus we are giving true value to our customer. In target costing most of the cost reduction is achieved during the design stage of the product. In what I call target budgeting the product is already mature and in manufacturing. Thus we will utilize kaizens and/or VE to reduce cost while in the manufacturing stage—what some call kaizen costing."

"I agree, Liz, but what is your point in using this target cost concept in budgeting?"

"By definition, standard costs are those carefully predetermined costs that are usually expressed on a per unit basis and are costs that should be attained. Standard costs are what we use in OPS budgets today. And I would ask the question, should last year's most representative actual cost be what we want to attain this year? I think not! If we do not have stretch goals, Mark, if all we want to achieve in our budget is last year's most representative actual or average cost or standard, we will slip backward in our Lean progress. The objective of all aspects of this LCMS—the Lean P&L, the Lean metrics, as well as target budgeting—is to keep us striving to reach our Lean goals and not slip backward to where we began, and also not to have the behavior changes we have created slip backward. The LCMS is the glue that holds all we have accomplished to date in place, and ensures that we do not regress. With its emphasis away from just financials and its emphasis on changing behavior, it is the inner combustion engine of our Lean program, which will help propel us forward and keep us moving forward.

"So my point is that, in planning our revenues and aggressively targeting our costs, we are also planning our profits. We will no longer be pushing this operating budget up and down between OPS and management for six months every year. We will be pulling from the customer! Management will determine up front what profit margin they desire for the year. That might be 8, 10, or 12 percent—whatever. If we know our revenues, and have scrutinized them well, and we know what we want for profit, then what is left in between is our target cost. This is our *go-get* or stretch goal if we want to make the desired profit. There will be no guessing as to whether or not we make a profit. We cannot keep asking people to just take 5 or 10 percent out of the costs of whatever budget they propose. We need to give people the knowledge on how best to do this budgeting, and then the problem-solving skills to devise the plan to get to the target. And so the target budgeting process is *once and be done!*"

"But you and I have talked with everyone about the danger of creating dysfunctional behavior by giving someone a number and telling them to go get it!" Mark seemed upset.

"When management tells us the desired profit they want to achieve, and we know what the customer is willing to pay for the product and what the market will allow, then we do have an aggressive cost target to go-get. But this is not just any target, Mark. This will be a very aggressive cost target. Thus, we will be asking each family to give us a plan for how they will achieve the desired target cost. There will be no games. This is more MBV! This target is not a standard that you can manipulate, then announce at year-end that you met the standard. Each month we will bring in each family and go through its budget, looking at actual cost this month compared to actual cost last month and then to year-end targets. And they must periodically update and communicate to us their detailed plan to reach any aggressive target. This is not a plan that you set to standard in May, and then go away for the remainder of the year! This *profit plan* will become our operating budget."

"Great! So how do we implement this?"

"I would suggest we have an all-day meeting with just you and me and the managers of the product families, for starters."

"OK, let's do it! I will set up a meeting for Friday, and we will jump-start this new target budget—or profit management process as you call it. This is truly a new management paradigm! I believe I will tell Jim that he is welcome to drop in toward the end of the meeting, to see the finished product. He has already given me a target percent of revenue that he wants to achieve in profits for next year for the A-team. I want to see him buy into what the families develop as an operating budget, and see how this all works."

On Friday morning, four of us got together in a conference room in the admin building. Together with Mark and myself were Mindy, product family manager for the A-team, and Tim, the product family manager for our second of five product families, the Tiger team. Like Mindy, Tim was very knowledgeable of the business. Unlike her, Tim was an engineer with knowledge of, but little experience in, Lean manufacturing. Both Mindy and Tim had no formal financial education or training. Tim, unlike Mindy, did have some experience in what I would call the traditional budgeting process.

"Hi, everyone! We have come together today to jump-start a pilot for the Lean target budget—another piece of the LCMS that will ensure that all the great accomplishments we have achieved to date will stick with us and we will continue to improve. Mark and I want to thank you for giving up your day in operations to work with us on this new budgeting model."

"I think we need to start with some basic definitions of what we have today, why it is detrimental to our Lean initiatives, and what we want to have in the near future," said Mark.

"I agree, Mark, and I will start with just that. I have brought with me a copy of last month's traditional income statement, which was distributed to some, but never all, persons in the business. Just look at the length and detail of this financial statement! And note that it is in what we call standard absorption cost, and includes a lot of what I would call financial gibberish. Mindy and Tim, have either of you heard of the terms *standard cost* and *absorption cost*?"

"I have," said Tim, and Mindy agreed that it was what, from her perspective, management used to beat OPS every month. "Standard cost, to me, is what the accounting department develops each year, and also what I believe is the number in our budget that we must not exceed. And I have heard that it is a calculation that is essentially just last year's most representative actual or average cost and, of course, carefully and mysteriously calculated by accounting." Tim smiled mischievously.

"It is most interesting to me, Tim, to hear your honest opinion of what you believe standard cost is. And you are correct! From a purely financial viewpoint, and what they teach accounting students at the university and in textbooks, standard cost is a traditional financial accounting system that was designed in the 1930s as an accounting system to calculate the cost of inventory and COGS, but is used today also as a management accounting system. It is designed to allocate indirect costs, which are everything other than material and direct labor, to individual products (something we don't do in Lean) based on labor hours or machine hours. This standard cost system thus encourages managers like you to produce unneeded products in order to absorb and spread overhead over a larger number of products, and thus minimize cost per unit by essentially keeping all of the people and machines busy making more product than you know the customer wants or more than you know you can sell."

"I believe we need to also review the differences between cost accounting and cost management, Liz," Mark added. "We have discussed this with our CEO and CFO, and they both have challenged this distinction many times. I still do not believe they understand the dissimilarities between cost accounting and cost management completely. I believe it would be good to discuss this with Mindy and Tim, also."

"Good point, Mark. Cost accounting is what I have just spoken of: allocating costs utilizing some one denominator with the purpose of valuing every part in inventory as well as calculating cost of goods sold. Cost accounting is very narrow in both scope and purpose. So I would describe cost accounting as standards, transactions (what you may hear as debits and credits), and what Mark would call command and control. That is to say, accountants try to command and control the business with this cost system. Cost management, on the other hand, has as its objective a well-organized, cross-functional team approach to cost reduction and cost control. Cost management starts at the design stage with QFD and VE. In cost management we attempt to trace or track costs such that all indirect costs (those costs that must be allocated in a financial cost accounting system) become direct costs that can be managed and acted upon by all functional persons, not just the accountants. After all, cost reduction should be everyone's job, not just an accountant's job."

"I understand," said Tim. "And I think, as an engineer, I know what target cost is all about. So how is target costing different from cost accounting and cost management?"

"Target costing, Tim and Mindy, is a cost management system! It will become our Lean cost management system. I would call it a *strategic* cost management process for reducing all costs in a family or in groups of similar products or services, and it focuses not on debits and credits and transactions but rather on the design of the products in the planning stage. So target cost is a planning tool for cost reduction, which involves all functional departments, from accounting to engineering to manufacturing to marketing and sales, and procurement. It targets value for the customer in a product that we know the customer will buy as well as targeting a plan for profits for the corporation. It is not a stand-alone practice performed by the accounting department as is standard costing. Rather it is a collection of multidisciplined people and methodologies such as QFD and VE that link functions such as engineering, marketing, quality, manufacturing, purchasing, and so on into a *system* to manage, control, and reduce costs."

"I get it, Liz," said Tim. "To hit the standard cost number, as you have said all along, we in OPS just have to stop endorsing overtime, or stop all training and other such discretionary expenses, or stop capital spending, or move some operations overseas. This is essentially what Detroit has done for decades, and look what has happened to those businesses? Cost management, or target costing, involves not just having a number to hit but also a plan to get there. I remember having fun with this at another sister company of this corporation. We all had to come to management, as the executives put it, in a monthly meeting and give our updated plans on how we were going to hit our year-end target costs. One day the Red Cross was in the building on a blood drive. Every family member came to the meeting that day with small buttons on their shirts that said, 'I GAVE.' Management and all of us laughed and laughed, and then management pretended to turn up the thermostat in the room and told us that we would give even more!"

"That is too funny, Tim," said Mark. "I had not heard that story. And you are right when you say that the budgeting process that utilizes target costing is not your traditional accounting department-focused budgeting exercise. You and Mindy will be collecting data and costs in your Lean P&L, Lean metrics, and value stream maps to help manage and plan your family's business, but you will also be striving to stretch to meet cost targets in an operations budget. Mindy and Tim, have you ever seen this:

TARGET COST = MARKET PRICE less TARGET PROFIT

"This tells us that cost is equal to the market-driven and customer-agreed-upon price minus a profit that the corporation desires to achieve. And we do something else that the accountants rarely do in cost accounting (I am just winding up Liz-san here) but that we engineers always do: we look at this target cost by life cycles. Those life cycles are more than just manufacturing costs. They also include the costs of R&D, sales and distribution, services and support, and disposal or recycling. How often, in a traditional cost system originating in accounting, do you see all of these life cycle costs included in the cost of the product? Think of the big tire manufacturing companies or the companies that make sodas or drinks and package them in aluminum cans or plastic bottles. Do they cost the product so that disposal or recycling costs are included? I doubt it." Mark was right-on from my perspective.

"I can think of a large consumer products business that introduced a product, a new drink, into the marketplace a few years back. This product was the brainstorm of an R&D person who was paid a bonus based on the number of new products he introduced each year. In the night courses I teach, I always ask my students the question, what is wrong with this picture?"

"I know!" said Mindy. "He did not ask the customer if he wanted this new drink, and I bet he never asked the customer what he would be willing to pay for this drink!"

"You are correct, Mindy. The business also bought new capital equipment to make this new drink, and then they had to actually produce a lot of the drink, bottle it, distribute it to test markets, and advertise it—with a big emphasis on advertising! Think of all the money they spent on a product that they did not even know the customer would want or buy."

"So what happened?" asked Tim and Mark.

"The new drink was a total flop, customers did not buy it, and the company discontinued it before it even became a new product. But think of the waste! I happened to be in a Lean conference down south one day when

I discovered that the sales team for this new product was in a conference room right next door. Even the salespeople were complaining about this product. And I certainly did not blame them. It is for sure that this company does not utilize target costing in any way. Someone had decided on his own volition that the customer wanted this product, so manufacturing made it, and then told marketing, 'We made it. Now you sell it!' So, what might be the solution to this problem, Mindy or Tim?"

"For starters, they need to incorporate QFD into the planning process, and then target cost to include all of the life cycle costs. But most of all, if they are going to pay bonuses to R&D people, they need to pay those bonuses based on new products the customer wants!" Tim said.

"So I presume this is what you had in mind, Liz, when you told us that target cost is a strategic control. And I guess you might call this target cost an *allowable* cost. If it does not include all of the costs of all of the life cycles, and it has not been designed to cost, then the new product introduction will not be allowed to continue."

"You have used a key phrase, Mark, designed to cost. We first ask the customer if he wants this product and what he would be willing to pay for this product. Then we ask management what they want as a profit for this product. What is left in between is the target cost, and we keep redesigning the new product until we get the cost to what the customer or market will allow. If we cannot do that, then we scrap the introduction of that new product. We never calculate a sales price by taking our standard costs for material, labor, and overhead and then adding a markup for profit. That is a sure way to failure! Price determines cost—never does cost determine price. So target costing for sure is a Lean cost management system as it makes people rethink the entire marketing and costing processes, as well as forcing them to reconsider what creates value for the customer. That consumer products company with the new drink did have a five-gate process for the introduction of new products, but the customer was not to be found in any of the five gates! Thus we see that target costing is also a cultural change away from the traditional department-focused accounting systems to process-focused, functionally integrated teams for cost management, with the voice of the customer always in mind."

"Also, I believe it is not enough to just launch cost reduction programs as most businesses do today. How can any one functional group, such as operations people, take cost out of a mature product if they do not understand how cost gets into the product? The costing process, in the past, has been a secret accountants have kept close to their vests and called a standard cost system. But then, too, we do not want Lone Rangers in any functional area performing point improvements just once a year. We want continuous, small, systematic, and holistic approaches to cost reduction. And no drive-by kaizens! They create incongruent activities and myopic behavior. I learned this from you, Liz." Mark proudly voiced his acceptance.

"So salespeople with consensus from the customer will determine the revenues, and management will determine the profit target. Then operations will set the targets for costs in their families, with a plan to achieve those most aggressive targets. Top management will do the same target budgeting for the SG&A (sales, general, and administrative) costs for which they are responsible—those 5 to 10 percent of costs that reside outside of the families. Then we just pull all of these family target budgets together with the SG&A management target budget, and we have a profit plan that will not fail. Thus, this is both a top-down and bottom-up budgeting process. We have planned both costs and profits, and now we have a win-win for the customer and the business. We will need to make sure that the sales team is being realistic, though, when they forecast number of units or number of pounds for the revenue calculation." I wanted to make sure that Mark, who visits our customers frequently, was on board with rationalizing the sales forecast.

"So I understand that target cost is a planning process, and we use that planning process to create a budget. But how will this actually work, Liz?"

"We begin with carefully calculated revenues, and I am not being facetious here. Mark has already spent time with the sales team and the customers, and has refined the revenues by month for us. Then Jim has communicated what he believes would be a realistic profit that he would like to see for the business as a whole for next year. Thus what is left in between the revenue and target profit is our target cost. Now, as material is the highest percentage of the cost of the business and each product family is a separate line of business with a unique material component, Jim may challenge each product family based on what he believes is realistic for that particular family's business or commodity. Because he knows the entire business well, especially commodity raw material prices, he may feel that the A-team is capable of reducing costs by 10 percent, but that another product family has a more difficult situation in the marketplace and thus he may allow them additional target costs. Once Jim has determined who gets what in the way of buckets of target dollars, then each family can spread its bucket of target dollars over its material and processing costs in any way they see fit. Thus, this target budgeting process is truly both top-down as well as bottom-up.

"We will take our Lean P&L as a model, and utilize that to make our target operating budget. We may want to explode the material line in the P&L to aid us in calculating what we need, to include quantity and price, from our suppliers. But when we are finished, we will have a plan to make the desired profit from the revenues that the sales team has firmly established by month. In other words, we have the revenues plugged into our Lean P&L target budget model on the top line. We know what our CEO wants for a percent profit for any given family, and we plug that gross profit percentage into the appropriate line at the bottom of the Lean P&L target budget model. What is then left between the revenue line and the gross profit line is our target cost, and we are free to spread our bucket of target cost dollars given to us by management over our material and processing, or actionable direct costs, in our family in any way we wish. To guide us overall, as we plug in costs for each line item, the total bucket of target costs in the bottom line is diminished by what we have already spent, so we always know what we have left to spend. Top management will accept any budget we create as long as we have a plan to get to each aggressive cost target, and we do not exceed our total bucket of target dollars for any given month. The customer gets what he wants at the price he wants, the CEO and top management get the profit they want, and we have a win-win situation for everyone in a short period of time—definitely less than the six-month budget process time we have now, and with no games. We know our revenues, we plan our profits, and thus we plan our costs. Once and be done!"

"I see how this works now, Liz," said Mark. "The concept of target costing as we engineers know it is for what I would call feasibility studies for new products. However, you have taken the concept and applied it to the Lean P&L to create a template for a feasible target cost operating budget for existing mature products. I like this! I especially like it because there is no standard cost involved. Your point was a good one, Liz, when you stated that we need to lock off this standard cost system to everyone outside of accounting. The standards for material, labor, and overhead should never be used in formulating a budget if we truly want to make the business much, much better year after year."

"OK, so if we are all in agreement, let's get started first with the A-team target budget for next year. Mark and the sales guys have already been out to visit our customers, and they have refined the revenues by month. I have plugged those revenue numbers into the model. Jim has made it known that the corporation desires a specific percent gross profit for the A-team product line, and that has been plugged into the model. So now we know how much we have in target costs to spread over our actionable costs each month. We can explode the line item for material costs, and we can explode any other line items we wish. We always know what we have left to spend at any given time in the budgeting process. If the business is large and we cannot handle all of this detail in a spreadsheet model, we can place this entire target budget in an Access model with drop-down boxes for people to enter the targets by part. There is no need to purchase any expensive software program—not here and not ever in Lean. Because some of our target costs will be very aggressive, we will need a plan for each very aggressive line item target, informing management of the process we propose to utilize to get to that target. So let's get busy!"

16

Town Hall Meeting—One Year Later

If you have put the Lean cost management system (including the Lean P&L and the Lean metrics) in place before the implementation of the Lean principles of value and flow in the plant, then you do not have the human behavior and culture in place to make sure your Lean improvements will keep happening instead of sliding back to where you began. Remember! Value and flow come first, then the Lean cost management system.

The town hall meeting this year-end was a celebration! It was quite different from the town hall one year earlier in which people had been lined up and then mown down with machine guns. No layoffs this year-end! People were happy! The members of the first product family, the A-team, were all dressed as characters from the early 1900s, such as Henry Ford, Tom Edison, and the Wright Brothers. They were demonstrating that they were the *innovators*. Their product family had been in place for eleven months. Members of the Tiger team, equally jubilant and in existence for six months, were dressed in the joyous fashion of the Chinese New Year. These two teams represented a little more than half of our business. Two smaller product line teams had been in place since September, and a family for a brand-new product line had just been created. Everyone brought a food dish, and all the dishes were displayed on tables in an area of the plant where 12 percent capacity had been created in the past year. There were shirts and hats for the celebration, and everyone was enjoying plates of food. No, there were no "hot" and "cold" lines for the food. Everyone just jumped in, chattering away with others.

Jim had grown professionally throughout the year, and had demonstrated with his actions and words that he truly had earned a Level 7 leadership position. He was insisting that problems be solved in a scientific manner, and that his management team interact with little or no conflict—only positive and constructive discussions on any disagreements. At 6 o'clock each morning, Jim could be seen parking his beemer in a far corner of the employee parking lot, walking a great distance to the plant, and stopping to speak with first-shift workers along the way. Then he walked slowly through the plant, stopping to ask every shop associate about his job and how "the work" was working. And he did the same thing in reverse at 6 o'clock every evening on the second shift. Members of the management team realized now that it is not through the creation of spreadsheets for allocation of costs nor through reading huge computer-generated reports at 2:00 a.m., but only through direct observation of, and data gathered on the shop floor that real improvements take place.

The CFO was even less approachable than she had been a year ago. Word on the street was that she had been offered a job in the Corporate Accounting (Core Finance) group, which was located in another state. She did demonstrate talent for Leaning the accounting functions such as A/P and A/R. However, she refused the offer. Word was that she did not want to relocate this late in her career. So Nancy was retiring, and Carol was temporarily taking her place until Jim and I could decide if this was the right choice or if we should bring in someone with Lean financial experience, preferably a CPA or CMA with a Six Sigma Black Belt.

Helen had come a long way in accepting change. She and I now needed to work on a methodology for actually implementing the systems of merit HR model. I had been giving this a lot of thought lately, especially in light of Nancy's decision. On what plateau on the systems of merit HR model did Nancy currently reside? Was she at Level 2 where she only wanted a paycheck? I think not. Was she at Level 3 where she wanted to fight everyone who believed in change and the LCMS, refusing to participate in the planning of both? I'm not sure. I am fairly certain she was at Level 4, the "command and control" plateau. But she also wanted to be "Queen." So perhaps that placed her at Level 5 where one puts oneself above the business and has nothing in mind other than one's own personal objectives. She definitely was not at Level 6, the teacher, nor at Level 7, the leader. So what will be the process, going forward, to identify on what plateau a person resides, and how do we place this person (and other persons) with a mentor to get them to the next level? This would be a huge undertaking for Helen and me and others in the business who wanted to help with this.

Mary had graduated in early December, but was going to wait for her classmates in order to participate in the big graduation ceremony in May.

As of January 2, she would be an associate of this company and a member of the newest product line family. She was uncertain at the time as to whether she wanted to sit for the CPA or CMA license exam—or perhaps she would do both! She is an overachiever! Then, too, Mark had plans to help her through the various Six Sigma belts to earn her Black Belt. Perhaps Mary will be our next controller or CFO!

In the beginning, we had too much inefficiency, too much resistance to change, too many aborted reorganizations with Lean tools only, and we were out of cash, living day to day on loans from the bank and the state. Our business philosophy had been short-term: increase profits by whatever financial reengineering or game-playing our associates (both accounting and OPS) chose to execute, hide inventory (along with defects and problems) under miles and piles of the stuff in buildings and trailers all over the plant, blame others for your mistakes or do not admit mistakes, and do whatever you have to do to get yourself promoted. This had all changed in the course of one year. Jim told the group of employees gathered in the now available space on the plant floor where capacity had been created that we had cut our inventory in half, making much cash available to us. He told everyone that we would use the cash to buy a small overseas supplier and have that supplier put a warehouse in this country, and that would better our metric of inventory turns. As expected, Jim did not show any specific numbers, but he did show visuals that displayed trends and painted a stunning picture for the business as a whole. The employees loved it! There was also a plan, which Jim communicated to our associates, for where we would be going as a company next year. We had a new customer, and that new customer and new product line were in a new market for us. As for the continuation of our Lean principles, Jim told the group that we would soon be starting the implementation of "pulling" from the customer, the fourth of the five Lean principles, and that would begin with the A-team, then the other teams would follow during the course of the year.

The best news of all, for me, came in the form of a letter from my sister, the nurse, back home. Inside of the letter was a small newspaper clipping that told the story of my mother receiving an award for the most hours donated by any volunteer in the history of the hospital. But that was not all! The newspaper story went on to say that my mother had been named to the Board of Directors of the hospital for her "innovative and creative cost containment ideas." So, you see, Henry Ford was correct when he said that the Lean methodology, as we know it today, is just "common sense," and the principles apply to any industry.

"So what have we learned this year, Liz-san?" Mark asked as we sat off to the side and observed the jubilation.

"We learned that one needs to implement value and flow first—before all else, Mark-san, and I learned (thanks to you) that leaders with passion trump charismatic leaders any day!"

"Yes, emotions trump commands. Even I see that every day with the dogs I train. And responsibility is an emotion, Liz, and cannot be commanded as you financial people might believe."

"Hey! Hey! I have always understood that command and control and police work do not belong in a business, especially in accounting. But then you engineers must concede some things also. You engineers are traditionally the Lean leader. You guys believe that change is merely a technical problem to be solved by rearranging the furniture, redesigning the plant floor, and creating flow and pull. But it is really a behavioral problem, Mark, created by a 1930s financial accounting system. So who, I ask, would be better to lead this Lean transformation than the finance guy or gal with a CPA or CMA and a Six Sigma Black Belt? 'Why?' you ask. Because we financial persons touch every corner of, and every person in, the business and the Lean change is first and foremost about people."

"Well, Liz, you have taught me that there is no one recipe for Lean or Lean accounting. There is no Step 1, Step 2, add one-half cup of empowerment and a dash of training. The new management paradigm is all about managing behavior, where empowerment means earning the trust of people, which means that you allow employees to vent now and then."

"I thought I was the queen of visuals, Mark, but your wheelbarrow escapade taught me a lesson in how to make a point so that people will take notice and start to make changes. Trends do tell a story, but wheelbarrows do it even better!"

"Jim told me about your study that gave reasons for failure in Lean implementations. Those reasons gave me food for thought, Liz. Lack of leadership, your first reason for failure, could certainly sink the ship. We usually have too many managers and no true leaders. Then lack of passion is another very good reason why businesses fail at Lean. We who lead the transformation need to be passionate about what we do, and make that passion contagious. Then, too, having values and a vision are what so many corporations do not have today in their businesses. Jack Welch preached this at GE, and he and his wife still preach this in speaking engagements today. But I must agree with you, Liz, that accounting and HR truly are the biggest limitations to a successful Lean implementation, and probably because they do touch every single associate."

"Yes, Mark, if you want to change the culture to a Lean culture, you must first change people's behavior, and to accomplish changes in behavior we must first change the accounting system by which we manage the business as well as the metrics by which we measure the progress in the new Lean business."

"Liz, I believe that every person on our Steering Team now realizes that in the past, in our first attempt at Lean, we were organized in departments and we were implementing tools. Today we have success because we have created an organization that is a system of linking cost and operations in each cross-functional product family. I also think, looking back on some of the sessions our Steering Team had, that we should have made problem solving a core value. Our customers surely want products that solve their problems. Universities have not been able to teach this, Liz, you are correct. Most graduates we get here are book smart, but they cannot solve problems. Perhaps we should teach, as part of our Learning Kaizens, the Six Sigma DMAIC process or the PDCA process and open a program for our people to earn one of the Six Sigma belts. Six Sigma and Lean do work together very well in business today: Six Sigma for quality, and Lean for speed to market. That is an awesome combo!"

"Yes, Mark, we cannot keep saying as we do with our kids, 'Clean your room!' or 'Take 10 percent out of the cost of your family budget!' We need to give people the knowledge on how to do this, and that includes teaching both the Lean principles as well as problem-solving skills."

"We should get to work on that dashboard we promised Jim a long time ago, too, Liz. With the Lean P&L and its trends, the Lean metrics, and all of the data in the value stream maps, we can do away with all of the PowerPoint decks that crop up in meetings everywhere. Dashboards trump decks! Then I know that you and Helen are working on our other big limitation, our HR system. But I think the two of you are going to need more help with this, Liz. We must carefully define the characteristics of each plateau, and then pair every associate with an associate, or mentor, at the next plateau. Even though we have abolished so much bad behavior by eliminating the traditional standard cost system and replacing it with the LCMS in the product families, which represent 90 percent of the cost of our business, are we not still rewarding some 10 percent of the people who sit outside of the product families for bad behavior with the HR system we currently have? Have we really changed behavior in all of our associates? This really came to my attention when we were working on the target budget with Mindy and Tim. We pay bonuses to the marketing people to push products the customer does not want, and then we pay salespeople bonuses to beat the plan they themselves created. We pay R&D people bonuses based on the number of new products they develop—forget about whether or not the product is something the customer wants. Then we pay HR to pigeonhole people based on personality tests, and create fear in the company. But the biggest blunder of all is paying our top management executives bonuses based on their ability to manipulate the traditional financial statements! Should we rather be rewarding these execs for meeting the targets in the new Lean metrics? Perhaps we should revisit our Learning Kaizen concept, include problem-solving skills training, and create more of these Learning Kaizens strictly for those associates who reside outside of the product families. After all, Lean is not just for manufacturing!"

As I sit alone tonight at home, I am sketching a flowchart to help me reflect on what has happened in this business over the past year. I believe we began this journey with a long-term vision and values that align with our customers. We then took the first three Lean principles—not just the Lean tools—embraced them, and moved to a new organizational structure where dedicated cross-functional product families replaced functional departmental silos. Thus, we became a decentralized company with less bureaucracy, allowing information to flow alongside material. Next we introduced the value stream map as a tool not just for the shop but also for top management, to aid them in the order of execution of events for continuous improvement, and developed the Learning Kaizen to make certain that the ideas that people presented in the value stream map were executed and put into action by those people on the shop floor who had the ideas. This created trust between shop workers and management. Then, to make certain that we did not slip backward, we introduced into the product families the LCMS, including the Lean P&L, the Lean target budget, and the new Lean metrics. And the timing of the implementation of these pieces of the LCMS—when and where we implement in relation to the principles of Lean—is so very important if we desire success (see Figure 16.1). All of this represents the way we are managing people today as opposed to the 'command and control' methodology of years past.

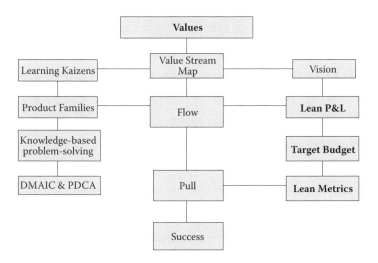

FIGURE 16.1
The timing of integration of LCMS into the Lean Principles.

We need to do more experimenting, however, to determine how our new Lean management system and LCMS will develop and grow for our particular business. We learned that changing behavior, and thus culture, is not accomplished with book clubs or by attending outside training seminars. We learn by doing, and making mistakes. We started with the hypothesis: "Structure drives behavior which creates a culture" and learned that, by *structure*, we are not suggesting just physical structure alone—functional silos or product families, centralized or decentralized. But structure also includes what we cannot see, and that is, how people learn, as well as how and why people behave as they do.

So this ending is really just a beginning! We have so much more to learn. However, it is not always the engineer who leads this change in behavior and thus culture. It is also the financial guy or gal. Just as Lean and Six Sigma professionals once sat far apart in a business and each believed that he or she had the only right answer, so the engineer and the accountant must no longer sit apart but join forces in continuously improving and leading the technical, as well as the behavioral, aspects of a Lean change management initiative.

17

The Role of Controller as Lean Leader

You, the controller, can be the Lean leader! You do not have to be charismatic or a Myers-Briggs extravert. Remember that behavior trumps personality! Leadership is all about demonstrating emotion and passion for what you believe in. It is about motivating people, not managing people, and not with the traditional carrot and stick but rather with the correct incentive, the correct game (perhaps challenge is a better choice of words), and the correct benefit for service. Leaders come from the ranks—those associates who know the business really well. Leaders do not try to control people but rather they create some ownership for people. They have a vision and a strategy, and they are not always an engineer by trade! Two of the very best Lean leaders I have ever seen in action were the finance guys who successfully led the Lean transformations at UTC Pratt & Whitney and Gorton Fish. And those Lean transformations were not just about numbers.

I have worn many "faces" as the Lean controller in this story. First and foremost, I am a business partner with the CEO, CFO, COO, HR, and plant managers. Equally important, I am a catalyst for change, and a champion of change. Then, too, I am an advocate for cost management—not cost accounting. Thus, I am a process manager, and not a transaction manager. I am a strategist, focused on the future, on innovation, and on performance. I am a steward of information—not just data. I am also a steward of effective, planned budgets utilizing target cost. I am a cash manager, but not a banker. Cash should be raised on the floor of the plant. I am a risk manager, including decisions on off-shoring. I am also a talent manager. Yes, I am a very unusual person for a controller. Specifically, I am a Lean controller.

My financial colleagues tell me that they know what Lean accounting is *not*, but that they are not sure what Lean accounting really is. Yes, it is true that Lean accounting is not a formal accounting statement. It is not (nor should it be) put together solely by accountants! It is not a CD-ROM or a piece of plug-and-play software, such as ABC Cost. It is also not a recipe that can be copied from someone's cookbook. And it does not cost individual products in inventory. Our financial cost system performs that tedious task.

So then what is Lean accounting? Lean accounting is a planning tool, created by OPS and for OPS. It is thus a methodology for costing the *processes* in the *value streams* for a *family* of products. Lean accounting ties together cost and operations in cross-functional families, and thus creates the correct behavior in a Lean entity, which in turn changes the culture of that business. Lean accounting is not only the combustion engine that propels us forward in changing the behavior of people, but it is also the rivets or structural fasteners which, if set correctly, will hold all parts of the Lean system together and impede us from falling apart in our Lean transformation. And it is just one piece of the Lean cost management system, which includes far more nonfinancial than financial information for managing the business.

If I could, I would add one more principle to the five original Lean principles, which are (1) define value, (2) map the value stream, (3) create flow, (4) pull from the customer, and (5) perfect. But, foremost, to the first principle, I would add, "Live your values!" Then the sixth principle I would add is (6) "Be *Humble*," for we can see by looking at history that every system, whether it was a business or a society or a civilization, that became arrogant has ceased to exist. One need only look back a very short time to see proof of arrogance annihilating a business.

To change the culture to a Lean culture, we first have to change people's behavior, and to change behavior we must change the 1930s financial accounting system we use to manage the business. And that new accounting system is a Lean cost management system.

Index